AF010872

VENUS
An Errant Twin

VENUS

AN ERRANT TWIN

ERIC BURGESS

COLUMBIA UNIVERSITY PRESS
NEW YORK

Library of Congress Cataloging in Publication Data

Burgess, Eric.
Venus, an errant twin.

Includes index.
1. Venus (Planet) 2. Space flight to Venus.
I. Title.
QB621.B87 1985 523.4'2 85-384
ISBN 0-231-05856-X (alk. paper)

Columbia University Press
New York Guildford, Surrey

Printed in the United States of America

c 10 9 8 7 6 5 4 3 2

Clothbound editions of Columbia University Press Books are
Smyth-sewn and printed on permanent and durable acid-free paper.

Book designed by Ken Venezio.

In memory of

Stephen Roy Burgess

"O my son, what dost thou not know?
What can I tell thee more?"

Contents

1 QUEEN OF THE HEAVENS

In February 1961 Soviet technicians at the secret launch base in Tyuratam swarmed around a large booster rocket, as they prepared for mankind's first mission to another planet. Tyuratam, the Soviet "Cosmodrome," is located in a broad and almost flat valley in the heart of what used to be the empire of Genghis Khan. Atop the big Sputnik 8 rocket was a small spacecraft named Venera 1. Its destination was the cloud-shrouded planet Venus, brightest object in Earth's skies after the Sun and Moon, and the planet which comes closest to Earth. Whenever the sky was clear of clouds in the days preceding the launch, engineers and scientists preparing the spacecraft were treated to the spectacle of Venus shining almost at its brightest above the semiarid desert. Then only 54 million miles from Earth, the intriguing planet beckoned with its brilliance as the western sky lost its sunset colors and deepened into the blackness of night.

The rocket engines of the big Soviet booster ignited on February 12. Ponderously the large rocket lifted its scientific payload, accelerating it to the more than 29,000 km/hr (18,000 mi/hr) needed to balance Earth's gravity and enter an orbit around the Earth. The Soviets were testing a method of launching interplanetary spacecraft from an orbiting satellite. At the appropriate time a command was issued and the spacecraft left the orbiter to start on its trail-blazing journey part way around the Sun for a rendezvous with Venus. Three months later, when Venus had passed between Earth and Sun and was a brilliant morning star, it was estimated that Venera 1 crossed the orbit of Venus and passed within 100,000 km (62,000 mi) of the cloud-shrouded planet. Unfortunately, the spacecraft did not return any useful scientific data from Venus. Venera 1 was not able to communicate with Earth from shortly after the time it had separated from the Sputnik 8 satellite launch platform. Telecommunications died when the spacecraft was about 75,000 km (47,000 mi) from Earth. Even the huge tracking station at Jodrell Bank, England, could not detect any signals from the spacecraft when a search was requested by Russian scientists.

The new era of planetary exploration continued in the summer of 1962 at the U.S. Atlantic Missile Range

based on Cape Canaveral, Florida, from which a generation of ballistic missiles had thundered out in success, or exploded in failure, across the Atlantic Ocean. The American Mariner 1 spacecraft was mounted within a protective fairing atop an Atlas-Agena booster, a peacetime application of one of the first intercontinental ballistic missiles. The Agena upper stage had been added to the Atlas ballistic missile to provide sufficient launch energy for the spacecraft to escape Earth's gravity and travel along an interplanetary trajectory. Again the interplanetary target was Venus. Again Venus was a brilliant evening star, dazzlingly beautiful in the sunset afterglow, taunting Earth's engineers and scientists to reach for her. And each evening when she had set beyond the western horizon, the darkness of the night sky was pierced by the eyes of the other planets watching the beings of Earth as they prepared to reach out into the solar system. Saturn turned a cold steady gaze on mankind from Capricornus in the southwest. Jupiter, rivaling Venus in brilliance, looked out from Aquarius in the south. Rising in the east as a companion of Aldebaran, eye of Taurus the bull, shone the baleful red eye of Mars. A waning half-moon also looked on over the stage being set for the first American attempt to unveil the mysteries of Earth's twin planet.

At 4:21 A.M. EDT on July 22, 1962, the three rocket engines of the Atlas ignited with a roar that could be heard across the flatlands of the launch center. The bright glare of the incandescent exhaust dimmed the light of moon, stars, and planets as the huge rocket rose majestically into the night sky. The big booster headed in the general direction of Mars; almost directly opposite to the horizon where Venus had set. It would use the spin of the Earth to add extra velocity as it headed out over the Atlantic Ocean; it would reach its interplanetary trajectory on the other side of the Earth.

At first all looked well for the gold-plated spacecraft. Then consternation! Radar and optical trackers warned that the Atlas was not on course; the huge rocket thundered toward the northeast instead of the southeast along a flight path that would carry it into shipping lanes of the North Atlantic and could even imperil areas of dense population on the East Coast. Steering commands were transmitted to the guidance system.

Hoping that the trajectory might be corrected and Mariner could be given a boost into its interplanetary orbit, the Range Safety Officer waited anxiously. But flaws in how the guidance equations were being processed within the booster's electronic controller made the huge rocket continue off course. At 4:26 A.M., after 293 seconds of flight and with only another six seconds to go before the scheduled time for the upper-stage Agena to separate from the Atlas, the Range Safety Officer pushed the destruct button. The huge rocket exploded into a ball of fire. But the Mariner spacecraft still transmitted plaintive telemetry signals as it arced along a ballistic trajectory over the Atlantic Ocean.

For just over a minute the Mariner sent engineering data over its radio link to the command center. And then it hit the water; the radio channel fell silent. The first American attempt to reach Venus had ended in a dismal failure. However, a second spacecraft had been built as insurance, and 37 days after the destruction of Mariner 1, Mariner 2 was poised ready for launching. Again an Atlas-Agena was checked out on its launch pad, with the spacecraft housed in a conical nose fairing on top of the huge two-stage rocket booster (figure 1.1). Countdown started at 5:37 P.M. EDT on August 26, 1962. Several technical delays occurred during the countdown, but at 1:53 A.M. the following morning, the Atlas engines flamed into action, and Mariner 2 sped on its way across the interplanetary void toward Venus.

Nearly four months later, on December 14, 1962, Mariner 2 crossed the orbit of Venus and flew by the planet at an altitude of 34,854 km (21,657 mi) above the hidden surface of the planet. During its 109-day interplanetary crossing, instruments on board Mariner 2 had monitored the environment of the unknown region of space between the orbits of the two planets and relayed back to Earth a constant stream of scientific data. It had discovered that a faint interplanetary magnetic field persists along the ecliptic plane, that the solar wind of charged particles—protons and electrons—flows from the Sun at velocities ranging from 320 to 770 km/sec (200 to 480 mi/sec), that the incidence of interplanetary dust is less near Venus than near Earth, and that the intensity of cosmic rays remained nearly constant along Mariner's flight path.

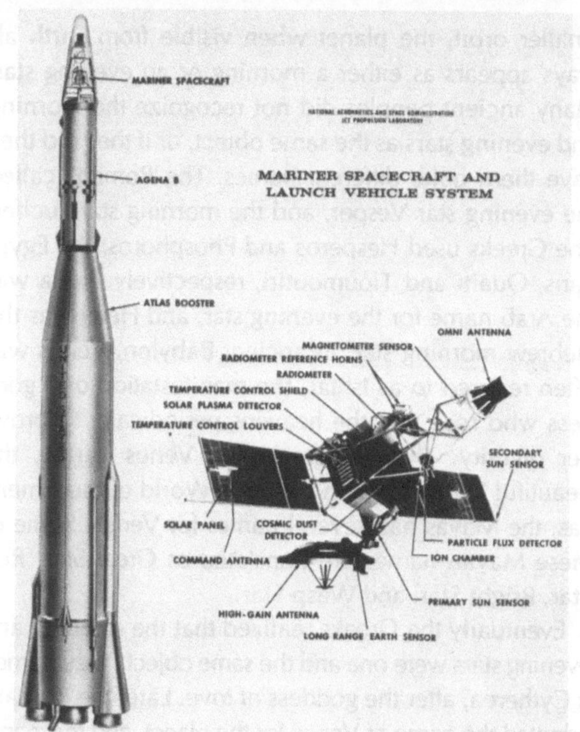

Figure 1.1. The first U.S. attempt to send a spacecraft to another planet, Venus, was with an early Mariner-class spacecraft using the Atlas-Agena as a launch vehicle. (Photo: NASA/JPL)

On December 14 the spacecraft had carried its instruments a thousand times closer to Venus than instruments based on Earth could ever come. For 17 minutes before and after the time of closest approach (periapsis) of the spacecraft to Venus, two temperature-measuring microwave radiometers scanned the planet on its night side, day side, and along the terminator, and confirmed that Venus is blanketed from view optically by cold, dense clouds in the planet's atmosphere. Also the radiometer data showed that the surface temperature must be about 430°C (800°F), a temperature at which lead and zinc would melt, were they present on the surface. This temperature was the same on day and night sides of the planet even though Venus rotates slowly on its axis and does not present only one face to the Sun. Other instruments discovered that Venus has neither a measurable magnetic field nor radiation belts, both so characteristic of the space environment near Earth.

The active exploration of Venus had begun. Over the next two decades the twin planet's veils of mystery would be lifted, and many questions that had puzzled astronomers and other scientists since the invention of the telescope would be answered. But the fascination of Venus reaches far back into human history. The brilliant planet has intrigued savants of numerous generations of many different peoples.

As they began to observe nature and to think about their surroundings, prehistoric men probably became aware that five luminous objects, which to the unaided human eye at first sight seem hardly different from the stars, change places and wander around the star sphere. These bright "stars" move along a narrow band passing through certain constellations of "fixed" stars, which are known as the constellations of the zodiac (derived from the Greek word for life, *zoe*). The names of these constellations [which with the exception of Libra (the scales) are all names of living things] have come to us without significant changes from the time of the ancient Greeks. Other ancient peoples—among them the Babylonians and the Chinese—also recognized twelve zodiacal constellations.

Venus is the brightest of the wandering stars (of course we now know they are planets). Sometimes Venus is bright enough to cast shadows on the Earth's surface, and it can be observed in the daytime if one knows exactly where to look for it. There is a story related by F. J. D. Arago, director of the Paris Observatory in the 1800s, that Napoleon Bonaparte was driving in his carriage to the Luxembourg to attend a fête given by the Directory in his honor. But the crowd assembled in the streets paid more attention to the heavens above the palace than to Napoleon or to the resplendent uniforms of his staff and entourage. When he inquired as to what claimed the attention of the people, he was informed that a star—Napoleon's star—was shining as a good omen for the Emperor in the noonday sky above the palace. The "star" was, of course, Venus.

Venus is brightest about one month before and after inferior conjunction, which is the time when Venus passes between the Earth and the Sun. Venus revolves around the Sun in an almost circular orbit in a period of 225 days at a mean distance of 108.16 million km (67.2 million mi) from the Sun (figure 1.2). Because of the Earth's own revolution around the Sun, the periods when Venus is visible at elongations and conjunctions in the skies of Earth, known as apparitions, repeat approximately every 584 days. Venus passes from greatest eastern elongation as an evening star to greatest western elongation as a morning star in about 140 days, and from a morning star back to an evening star in about 430 days. This frequent brilliance of the planet undoubtedly led to the importance with which Venus was regarded by many ancient observers of the skies.

However, many of these observers appeared to have been confused by the apparitions of Venus. Because the planet's orbit is closer to the Sun than Earth's is it cannot move far from the Sun in the sky as observed from Earth. Sometimes Venus precedes the rising Sun in the morning sky. At other times it sets after the Sun in the evening (figure 1.3). The planet cannot be seen throughout the night as can those planets which move in orbits outside that of Earth. As a result of Venus's smaller orbit, the planet when visible from Earth always appears as either a morning or an evening star. Many ancient peoples did not recognize the morning and evening stars as the same object, or if they did they gave them quite different names. The Romans called the evening star Vesper, and the morning star Lucifer. The Greeks used Hesperos and Phosphoros; the Egyptians, Quaiti and Tioumoutiri, respectively. Ruda was the Arab name for the evening star, and Helel was the Hebrew morning star. In ancient Babylon, Venus was often referred to as Ishtar, the manifestation of a goddess who rose into the heavens periodically to prove her divinity. The Chinese called Venus Tai-pe, the Beautiful White One. In the New World of the Americas, the Mayas had several names for Venus. Some of these Mayan names are translated as Great Star, Red Star, Bright Star, and Wasp Star.

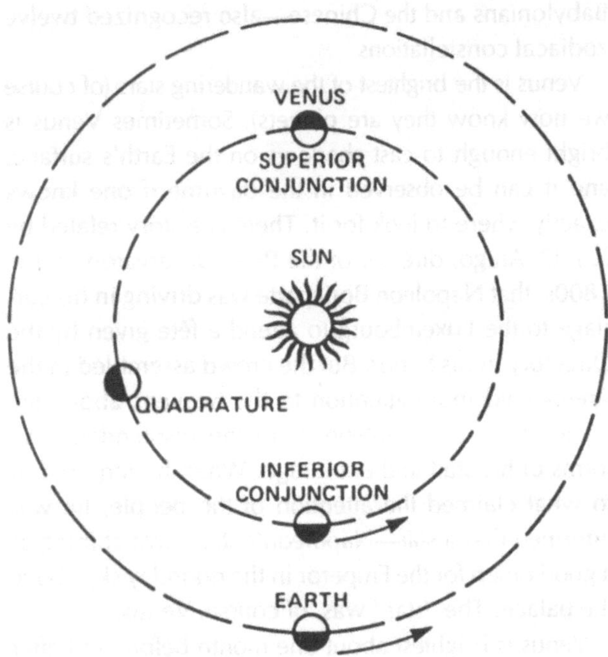

Figure 1.2. Venus orbits the Sun within Earth's orbit; its apparitions include inferior and superior conjunctions and eastern and western elongations.

Eventually the Greeks realized that the morning and evening stars were one and the same object. They named it Cytherea, after the goddess of love. Later the Romans adopted the name of Venus for the planet, and this name has persisted to the present day.

The Babylonians diligently recorded the apparitions of Venus, which was often referred to in the texts as Nin-dar-anna (Mistress of the Heavens). One ancient text, now preserved in the British Museum from King Ashurbanipal's library, describes how Venus appears in the eastern sky before sunrise and then disappears for three months before appearing again in the western sky after sunset. The text describes how, after staying in the western sky for just over eight months, Nin-dar-anna disappears for about a week before appearing again in the eastern sky for a period of just over eight months.

Evidence preserved in the available records from ancient Babylon suggests very strongly that Venus was being observed during the first period of Babylonian power. And this contrasts markedly with the much sparser records of observations of other planets. There is even some evidence that the Babylonians might have observed, in the clear atmosphere of their lands, the crescent shape of the planet when it is closest to Earth, something European scientists were not aware of until the time of Galileo. This derives from a text which has been interpreted as stating that "when Ishtar (Venus) at

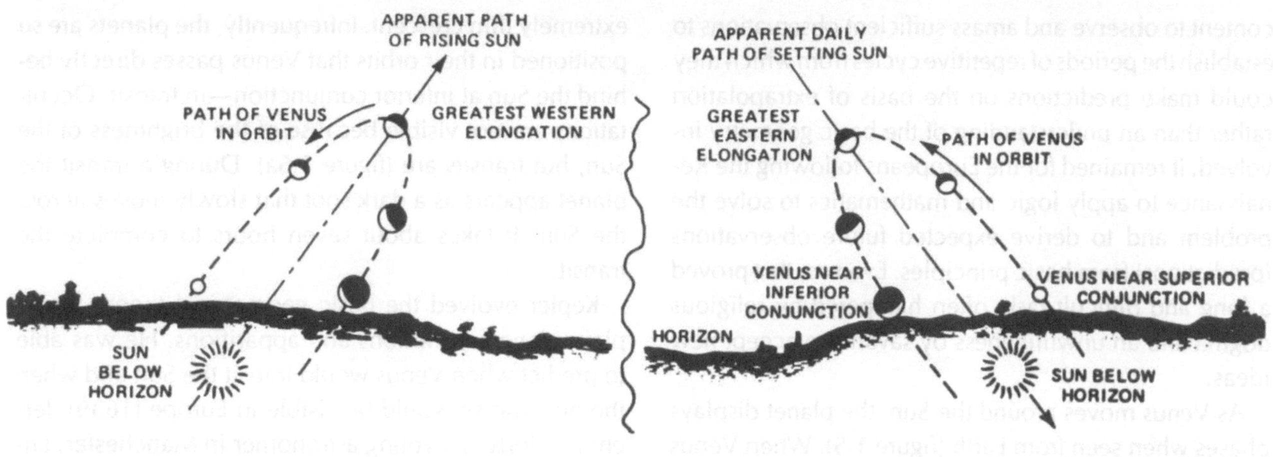

Figure 1.3. Venus can be observed from Earth only as a morning or an evening "star," rising a few hours before the Sun at western elongations, and setting a few hours after the Sun at Eastern elongations.

her right-hand horn approaches a star . . ." the signs are favorable, and vice versa with the other horn. The implication is that the Babylonians must have known of the horned crescent shape to mention the two horns in the text. However, "horn" might also have referred to the elongations either side of the Sun.

That the Babylonians gave such attention to Venus may have arisen because of the need of many ancient peoples to determine the beginning of the lunar month by the first observation of the thin crescent of the New Moon (figure 1.4). Venus would often appear in the same region of the sky, and patterns of observations would undoubtedly be noticed. The ancient priest-observers tried to find the periods of apparitions of the heavenly bodies which they could then use for predictions. More recently the Mayas paid particular attention to Venus and derived very accurate calendars for predicting the positions and apparitions of the planet. They set up temples and stone alignments to pinpoint where Venus would rise or set on the horizon at different periods in the cycles of Venus and the seasons of Earth.

But despite the accuracy of their observations none of the ancient civilizations appears to have attempted logical explanations for the cycles. The observers were

Figure 1.4. Venus became important to ancient peoples as they watched for the new moon in ceremonies to determine the start of the lunar month. Often they would see Venus and become aware of its cycles also. (Photo NASA/JPL)

content to observe and amass sufficient observations to establish the periods of repetitive cycles from which they could make predictions on the basis of extrapolation rather than an understanding of the basic geometry involved. It remained for the Europeans following the Renaissance to apply logic and mathematics to solve the problem and to derive expected future observations (predictions) from basic principles. Even so, this proved a long and difficult task, often hampered by religious dogma and an unwillingness by savants to accept new ideas.

As Venus moves around the Sun, the planet displays phases when seen from Earth (figure 1.5). When Venus is on the far side of the Sun it appears fully illuminated, like a full moon. As the planet swings between Earth and Sun, it shows a narrowing crescent phase. Its apparent size increases, but the amount of illuminated surface decreases. When Venus passes exactly between Earth and Sun at inferior conjunction it can be seen in transit across the face of the Sun.

However, the orbits of Venus and Earth are not in the same plane but are tilted about 3.5 degrees relative to each other. As a result, Venus normally passes through conjunction above or below the Sun and appears as an extremely thin crescent. Infrequently, the planets are so positioned in their orbits that Venus passes directly behind the Sun at inferior conjunction—in transit. Occultations are not visible because of the brightness of the Sun, but transits are (figure 1.6a). During a transit the planet appears as a dark spot that slowly moves across the Sun. It takes about seven hours to complete the transit.

Kepler evolved the basic geometrical theory to explain planetary motions and apparitions. He was able to predict when Venus would transit the Sun and when the next transit would be visible in Europe (1639). Jeremiah Horrox, a young astronomer in Manchester, England, was aware of Kepler's prediction. But he did not agree with Kepler's calculation, which predicted that the transit would begin at 8:08 A.M. in Manchester. Horrox recalculated the time as being 5:57 P.M. on November 24, 1639 (which was December 4 on the new Gregorian calendar introduced by Pope Gregory). He set up a telescope to project an image of the Sun on a screen and he watched expectantly for the predicted black spot of Venus. He was rewarded by being the first person to record seeing a transit of Venus. His observation was confirmed by his friend William Crabtree who at Horrox's request set up a telescope over his shop and saw the transit at 4:00 P.M. The accomplish-

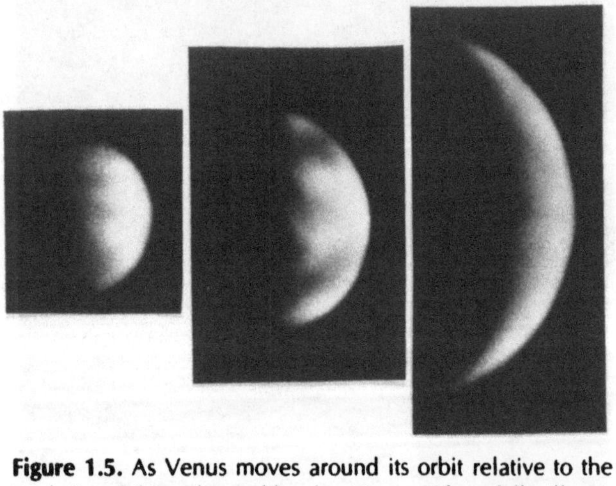

Figure 1.5. As Venus moves around its orbit relative to the Earth it exhibits phases like the Moon. When fully illuminated in the view from Earth, Venus is most distant from Earth and presents only a small disc for observation. When closest to Earth and presenting a much larger disc, the planet is, however, illuminated only as a thin crescent, and the side of the planet turned toward Earth is in darkness.

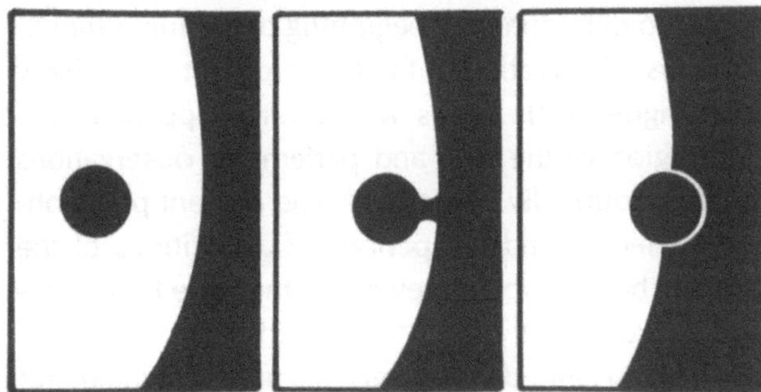

Figure 1.6. Sometimes when Venus passes through inferior conjunction it is close to a point on its orbit where the orbits of Earth and Venus intersect. Then a transit of Venus across the face of the Sun occurs and the planet is observed as a small black spot moving across the bright solar disc. Astronomers attempted to observe transits of Venus to determine the distance of the planet and of the Sun, but optical effects made it difficult to determine the exact time when the planet was at the edge of the solar disc.

ment is memorialized by a mural on one of the corridors of the Manchester Town Hall (figure 1.6b), and by a marble scroll placed on the monument at the west end of the nave of Westminster Abbey in London. The inscription describes Horrox as Curate of Hoole in Lancashire and, after listing his other accomplishments in astronomy, it states: "And predicted from his own observations, the Transit of Venus which was seen by himself and his friend William Crabtree on Sunday, the 24th November (O.S.) 1639."

Transits of Venus became very important to astronomers over the following century. A major problem faced by scientists was that of determining the true distance of Earth from the Sun, the mean value of which is now referred to as the astronomical unit. On this value all other calculated astronomical distances and sizes in the Solar System depended, and stellar distances too. In the 1670s the English astronomer Edmund Halley was sent to St. Helena in the South Atlantic to measure the positions of stars. While on this assignment he also observed a transit of Mercury across the face of the Sun and he attempted, by timing the transit, to derive the distance between Earth and Sun. Although unsuccessful in this effort, he reasoned that Venus would be a much better object for such an observation. He recommended observations of transits of Venus to determine the solar parallax to two orders of magnitude greater accuracy than hitherto possible.

Unfortunately, transits of Venus occur very rarely, in close pairs separated by more than a century. Halley predicted that the next pair of transits would occur on June 6, 1761, and June 3, 1769. In 1716 he emphasized the importance of observing these transits. However, they would be only partially visible in Europe. To achieve the purpose of measuring the astronomical unit it was necessary to measure the parallax of the Sun to a high order of precision—i.e., how much the position of the Sun (in this observation, of Venus) differs when observed from different places, thus giving a baseline and angles of a triangle to solve the lengths of the other sides. This could be accomplished, said Halley, by positioning observers at widely separated locations on Earth and accurately timing when the disc of Venus moved completely onto the solar disc, and when it touched the edge of the solar disc at egress. Differences in the times could then be used to compute the distance of

Figure 1.6(b). The first observation of a transit of Venus was made by Horrox and Crabtree. In this mural at Manchester Town Hall, England, Crabtree is depicted making his observation confirmed the observation made separately by Horrox. Copyright: City of Manchester.

Venus and the Sun trigonometrically by using the baselines of the various locations on Earth.

Although Halley died before the next pair of transits, his appeal met with an enthusiastic response from many French and English astronomers who spread an observing network over the world, from Siberia to India, and across the Pacific. Captain Cook's voyage into the South Pacific to prepare for the second transit of the pair led to the discovery of Tahiti. For this second transit the observing net ranged even further afield to California, Hudson Bay, Tahiti, and to Vardo within the Arctic circle. The French government even instructed its warships not to molest Captain Cook's ships on their expedition to the Pacific because they were on a scientific mission. Many astronomers suffered excruciating hardships and a few were killed by accident or died from strange diseases while on the transit expeditions.

But despite the enormous national efforts for that period of human history, the results were disappointing. The exact timing of the dark spot leaving the edge of the Sun at ingress proved impossible because of optical effects. As the small dark disc crossed the edge of the large bright disc, the black disk of Venus remained attached like a teardrop to the Sun's border (see figure 1.6), until the dark connecting thread suddenly broke and revealed the black dot separated from the edge of the Sun. Observers had to make estimates of when the true separation occurred and their times differed by tens of seconds among observers standing closely together with their instruments. A somewhat similar optical effect marred the timing of egress. Another major problem was that the geographical coordinates of many of the places of observation were very inaccurately known, especially the longitude. Nevertheless the distance of the Sun was determined more accurately than before, especially at the second transit of the pair, but not with the precision that Halley had expected.

These transits also provided important information about Venus itself. At the first transit, in 1761, the Russian astronomer M. V. Lomonosov observed that as Venus contacted the solar disc the edge of the bright disc appeared to extend toward the planet, almost as though the Sun were reaching out to pull the diminutive spot of the planet onto the disc. Also a faint halo surrounded Venus. Lomonosov correctly reasoned that these effects originated from the presence of an atmosphere surrounding Venus. Thirty years later the German astronomer Schroeter observed that, as Venus approached inferior conjunction, the horns of the thin illuminated crescent of the planet could be seen to extend beyond the geometry of a sphere and to reach around the dark limb of the planet (figure 1.7). This, too, showed that Venus has a substantial atmosphere, and confirmed Lomonosov's discovery.

But Venus remained a mysterious planet because of this atmosphere. Although Venus had been attentively studied for several centuries comparatively little was known about the planet even by the beginning of the space age in 1957. The major question was Venus's period of rotation. In 1788, after nine years of observations, Schroeter concluded that Venus took 23 hours 28 minutes to rotate on its axis, supporting the claim made over a century earlier by Cassini. But in 1890 Schiaparelli disputed all the earlier observations of Venus and claimed that the planet rotated synchronously with its revolution about the Sun, namely in 225 days. A great controversy raged for another 71 years with

Figure 1.7. When Venus is close to the Sun in the sky as seen from Earth, the horns of the crescent extend around the dark globe and show astronomers that the planet has an atmosphere.

claims of rotation periods ranging from as little as 12 hours to synchronous rotation in which one hemisphere of Venus would always face the Sun. Even the application of spectroscopic observations did not help. When observed spectroscopically the approaching limb of the planet should show a shift of spectral lines toward the violet end of the spectrum and the receding limb should show a shift toward the red. Attempts to measure these spectroscopic Doppler shifts by Richardson at Mount Wilson Observatory in 1956 showed that the shifts were so slight that the rotation period had to be greater than 24 hours. A rotation period similar to Earth would have shown up clearly in the spectroscopic observations.

Although it is generally accepted that Galileo discovered the phases of Venus when he first observed the planet through a telescope in the 1600s, the following centuries of telescopic study revealed little more about the planet because a blanket of bright yellowish clouds hides the surface at all times. Because of its phases Venus is not an easy planet to observe telescopically from Earth, and the problem is compounded by the planet's brilliant envelope of clouds. Astronomers stared at the planet through all types of filters attached to their telescopes but could only speculate about what might be beneath the impenetrable cloud layers and very little was definitively known about Venus before it could be explored by spacecraft.

All surface features of Venus are obscured by its pale yellow, opaque cloudy mask and the planet is featureless when observed visually or when photographed in visible or in infrared light. Only on images of the planet obtained in ultraviolet light are diffuse markings visible. These were attributed to materials in the clouds absorbing ultraviolet light.

Speculations about Venus had been rampant for centuries, particularly because of its featurelessness. For example, features had been observed on Mars and Jupiter to establish the periods of rotation of those planets. Schroeter claimed that he had seen a blunted horn to the crescent Venus, which represented an enormous mountain on the planet near its south pole. He also claimed to have observed irregularities on the terminator greater than those seen in our Moon, where they are caused by the presence of craters and mountains. Herschel disputed these "discoveries," but other astute observers such as Beer and Madler supported Schroeter. White patches on the poles were claimed to have been seen by Gruithuisen in 1813, and by others later; and the astronomer M. Quenisset claimed he had photographed them in 1908 at Juvissy, France. Another French astronomer, Trouvelot, claimed he had seen white spots on Venus in 1876 and 1877 that were similar to spots he had observed on Mars. Closer to our time, Lowell stated he had seen linear markings on Venus somewhat like those he claimed for Mars.

Astronomers had earlier reported seeing faint low-contrast markings when the planet was observed through a yellow filter. Some speculated that these markings represented a clearing of the clouds to provide a brief glimpse of the surface. Others thought that the visual markings represented meteorological patterns, such as groupings of clouds, formed by mountainous areas of the planet.

A major discovery occurred in 1957, when Boyer observed Venus in ultraviolet light and detected markings on the clouds. Continued observations showed that these markings (figure 1.8) varied in shape and appeared to rotate about the planet in a period of four days, but in a *retrograde* direction—i.e., opposite to the Earth's pattern and the general rotation of planets.

Another mystery that had intrigued astronomers for a long time was the ashen light of Venus, akin to the appearance of the "old moon in the young moon's arms" seen when Earthshine lights up the night side of the

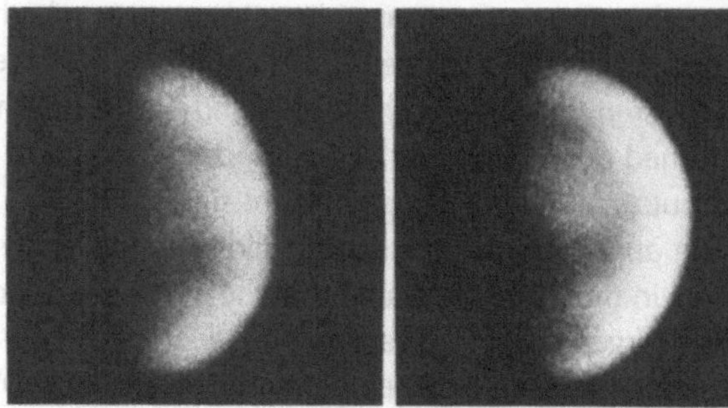

Figure 1.8. Photographs of Venus from earth taken in ultraviolet light reveal faint markings that cannot be seen in visible light observations.

crescent moon. The phenomenon seems to have been mentioned first by Derham in 1715. In 1721, Kirch, an assistant astronomer at the Berlin Academy of Sciences, also observed the ashen light.

In 1806 Schroeter stated he had observed the whole of the dark part of the disc visible as an ash-colored light. The same faint illumination was observed in January by Harding at Gottingen, and he saw it again the following month. Many other observers saw this illumination in subsequent years. While some astronomers attributed this faint glow on the dark side of Venus to Earthshine, others followed the lead of Arago and attributed it to auroral effects. Among the wilder speculations were that the planet glowed with intense heat or that its oceans were luminous. One astronomer claimed that the ashen light was brightest at times of maximum solar activity, which seemed to lend credence to the auroral hypothesis.

Yet the nature of the ashen light was still being disputed well into the space age and has not been generally accepted at the time of writing. In the 1930s several astronomers analyzed reported observations and concluded that the ashen light was not a real phenomenon. The Soviet astronomer Kozyrev made spectral measurements of the night side of Venus in the 1950s and detected emission lines of ionized gases. The data indicated that a Venus airglow was about 50 times as intense as the airglow of a terrestrial night. An important derivation from the data was that Venus would have to have an ionosphere to allow such an airglow to be generated by recombination of ions and electrons on the night side of the planet.

Another great controversy about Venus was whether or not it possessed a satellite. Alleged satellites were 'discovered' throughout the 1700s and 1800s but they all turned out to be stars. Venus is now accepted to be without a satellite of any significant size.

The application of the spectroscope to the study of Venus in the 1930s revealed that the atmosphere contained large amounts of carbon dioxide. Infrared observations indicated cloud temperatures of about −40° C (−40° F). When radio waves were detected from the planet in the new age of radio astronomy, speculation again ran to extremes. At short radio wavelengths the indicated temperatures were low, but at longer wavelengths a much higher temperature was indicated. The radio emanations could be interpreted in two quite different ways. They might originate from an extremely hot surface—hot enough, indeed, to melt lead and zinc. Or they might originate from ionized gases in the upper atmosphere. If the latter interpretation were correct the surface could be cool enough to support life.

In 1956 radar echoes were first bounced off Venus by scientists of the Lincoln Laboratory at MIT in an attempt to determine the astronomical unit more accurately. In subsequent years the experiment was repeated by scientists elsewhere. But it was not until the inferior conjunction of 1961 that the Lincoln Laboratory experiments produced echoes with characteristics showing that Venus rotates in a retrograde direction in a period of approximately 220 days. However, Soviet experimenters using similar equipment and frequencies measured a period of nine days. In the following year the experiments were repeated with radars of greater sensitivity. Carpenter and Goldstein of California Institute of Technology obtained signals that could be interpreted as a 1200-day direct rotation period or a 230-day retrograde rotation period. Soviet investigators had by this time also determined that the rotation period lay between 200 and 300 days in a retrograde direction. By 1964 the results were becoming more accurate. The period had been narrowed to 240 days, fairly close to the 243.1 days now accepted. However, the major mystery remained of how the surface of the planet as observed by radar could rotate in 243 days while cloud markings, which had been observed on photographs made by ultraviolet light, showed a rotation period of four days.

An intriguing aspect of the rotation period of Venus was that the same face of the planet would always be turned toward Earth at the time of inferior conjunction, when Venus is closest to the Earth. Later measurements of the rotation period showed, however, that this synchronism was not exact. Nevertheless, the long period of rotation in a retrograde direction to the motion in orbit leads to a solar day on Venus of 116.8 Earth days (figure 1.9). Statistical details of Venus are summarized in table 1.1.

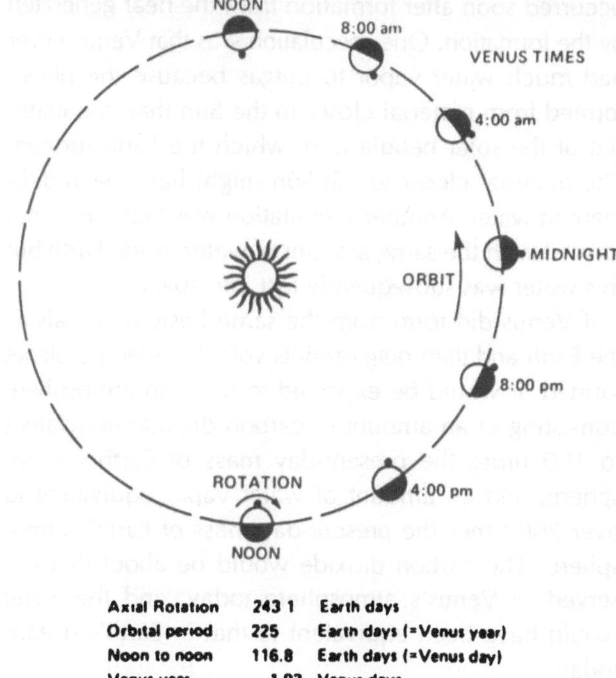

Figure 1.9. The combination of the slow retrograde rotation of Venus on its axis, and the revolution of the planet around the Sun, makes a day on Venus last 116.8 Earth days. A year on Venus is only 1.93 Venus days, which is 225 Earth days.

Of all the planets of the solar system Venus would seem to be the most promising subject for exploration. It is about the same size as Earth, it is the nearest planet to Earth, and it receives about the same amount of heat from the Sun. As Earth's twin, it would be expected to be a somewhat similar planet in other respects as well. But the secrets of Venus have yielded very slowly to scientific investigation and as they did, the revelations showed that Venus is significantly different from Earth in many important respects.

Before spacecraft explored Venus, speculations about the surface of the planet were wild. Some visualized a steamy swampy world, like Earth in the Paleozoic age, with great reptiles, tree ferns, and rain forests. Others imagined a dusty planet of stark mountains, sandblasted by planetwide dust storms. Others speculated that Venus had great oceans of carbonated water, while other speculations were of a planet covered by hot oil.

Astronomers, unable to see any surface details on Venus, attributed the obscuration to clouds in the atmosphere. The pale yellow clouds are so very reflective they return to space about 75 percent of the sunlight falling on them. Originally the clouds were thought to be of condensed water vapor like the clouds of Earth. However, repeated attempts to detect water vapor were unsuccessful.

The composition of the atmosphere of Venus was much in doubt until carbon dioxide was identified spectroscopically in 1932, when Dunham and Adams photographed the spectrum of Venus in a wavelength near 800 A searching for water vapor in the planet's atmosphere. They found no spectral lines for water but

Table 1.1. Venus's Vital Statistics

ORBITAL		
Mean distance from Sun	0.723	astronomical units
	108.2	million km
	67.2	million mi
Inclination of orbit to Earth's orbit	3.39	degrees
Orbital eccentricity	0.0068	
Sidereal period	224.7	Earth days
Mean synodic period	583.9	Earth days
Mean orbital velocity	35.05	km/sec
	21.78	mi/sec
Stellar magnitude when brightest	−4.4	
Angular diameter when closest to Earth	60.8	arcsec
Closest approach to Earth	41.5	million km
	25.7	million mi
PLANETARY		
Diameter (solid surface)	12,100.0	km
	7,519.0	mi
Mass	0.815	Earth masses
Density	5.26	gm/cc
Axial rotation period (retrograde)	243.1	Earth days
Rotation period, cloud tops (retrograde)	4	Earth days (approx)
Period of solar day	116.8	Earth days
Inclination of rotation axis	177	degrees
Surface gravity	888	cm/sec/sec
Surface atmospheric pressure	95	atmospheres
	9,616	kPa
	1,396	psi
Surface temperature	750	K (approx)
	480°	C (approx)
	900°	F (approx)

they did see two strange absorption bands near to wavelengths of 7820 and 7883 A. There was inadequate laboratory research to identify these strange bands, but later it was discovered through improved laboratory tests devised by the experimenters that the bands could be produced by carbon dioxide.

In 1947 Kuiper photographed other bands of carbon dioxide elsewhere in the Venus spectrum, and still more evidences of carbon dioxide were recorded in subsequent years of observations from Earth; however, these spectroscopic searches failed to reveal other major constituents of the dense atmosphere. Scientists assumed that by analogy with Earth there would be a large percentage of nitrogen present, but it was not possible to confirm this spectroscopically. Later, in 1961, traces were found of carbon monoxide. Although traces of water vapor were detected in the early 1960s by observations from high-altitude balloons, oxygen could not be clearly identified on Venus despite many attempts to do so.

Discovery of carbon dioxide in the atmosphere of Venus undoubtedly marked the first real step forward toward understanding that planet's atmosphere. But until spacecraft could penetrate the atmosphere and detect other gases there was no infallible way of telling how much of the atmosphere consisted of carbon dioxide. Although spectroscopy cannot identify the exact amount of carbon dioxide, it did suggest that this gas was the major constituent of Venus's atmosphere.

The amount of carbon dioxide is important because it determines how the microwave spectrum of the planet is interpreted. With very high percentages of carbon dioxide in the atmosphere of Venus, the microwave observations would permit as little as 0.5% water below the clouds. However, if another gas that did not readily absorb microwaves were present, the planet's atmosphere could have a greater percentage of water vapor and still satisfy the microwave data.

Carbon dioxide is also significant in our trying to understand the evolution of a planet's atmosphere. The atmospheres of both Venus and Earth are assumed to have originated from gases that were released from the interiors of the planets. Most of the outgassing may have occurred soon after formation from the heat generated by the formation. One speculation was that Venus never had much water vapor to outgas because the planet formed from material closer to the Sun than the material of the solar nebula from which the Earth formed. The material closer to the Sun might have been deficient in water. Another speculation was that the planet formed with the same amount of water as the Earth but this water was subsequently lost into space.

If Venus did form from the same basic materials as the Earth and then outgassed its volatiles after the planet formed, it would be expected to have an atmosphere consisting of an amount of carbon dioxide equivalent to 100 times the present-day mass of Earth's atmosphere, and an amount of water vapor equivalent to over 200 times the present-day mass of Earth's atmosphere. The carbon dioxide would be about that observed in Venus's atmosphere today, and the water would have been equivalent to that in Earth's oceans today.

The Earth holds its water in oceans because it is much cooler than Venus and there is a cold trap in its atmosphere below an ozone layer which strongly absorbs solar ultraviolet radiation. This trap prevents water vapor from rising to heights in Earth's atmosphere at which incoming solar ultraviolet radiation could dissociate the water molecules into oxygen and hydrogen, thereby allowing the lightweight hydrogen atoms to escape into space. Scientists speculated that if Earth were moved to the distance of Venus from the Sun the additional solar energy input would evaporate Earth's oceans, break the stratospheric cold trap, and allow hydrogen to leak into space. Earth would lose its oceans. This could have happened on Venus.

If the early atmosphere of Venus had been heavily laden with steam—because of the planet's proximity to the Sun and resultant higher temperature than Earth—the convective atmosphere could not have had a cold trap like that of Earth's atmosphere. Through dissociation of water vapor into hydrogen and oxygen, conditions on the planet could have changed enormously within about 30 million years. As much as 90 percent of the water could have been lost from Venus forever, the hydrogen escaping into space and the oxygen combining with surface rocks. But it is difficult to account for the disappearance of the remaining 10 percent.

Also, without there being running water on the surface to continually expose fresh rocks that could react with the oxygen, it is difficult to account for the disappearance of oxygen from the atmosphere. Moreover, scientists questioned what could be happening to the oxygen released into the atmosphere by the dissociation of carbon dioxide today. Observations from Earth had not shown appreciable quantities of oxygen in the planet's atmosphere. It was believed that the dissociation was occurring because of carbon monoxide (a product of such dissociation), the presence of which had been detected from Earth. In 1973 sulfuric acid was identified in the atmosphere of Venus. A possibility was that the oxygen was being removed by combining with sulfur to produce the acid.

On Venus, because of its higher surface temperature, reactions between rocks and the atmospheric gases would be expected to take place more vigorously than on Earth. However, the terrestrial surface is continually molded by running water to expose new surfaces on which reactions can occur. Since this process was thought to be unlikely on Venus because of the scarcity of water, scientists speculated that the surface of Venus might not have reached equilibrium with the atmosphere even to this day.

Before spacecraft penetrated the atmosphere and landed on Venus, inferences about the surface conditions on the planet had to be based on observations at other than optical wavelengths. Before the advent of radio astronomy all assumptions about the planet's surface were highly speculative. In 1956 radio waves at 3-cm wavelength were first detected from Venus and showed that the surface temperature was greater than 300° C (570° F). In subsequent years many other observations were made and the data appeared to cluster in three ways. Millimeter wavelengths showed a surface temperature of about 110° C (230° F), centimeter waves a temperature of 280° C (536° F), and decimeter waves a temperature of about 400° C (750° F). But it was not until 1968 that the really high surface temperature was generally recognized. By then radar and radio observations indicated a surface temperature of at least 480° C (900° F), and a surface atmospheric pressure was postulated of at least 90 times that of the Earth's atmosphere at sea level.

Nevertheless by the time of the first space missions to Venus scientists had determined very little about Earth's twin. The orbit of Venus was known and its ephemeris well established. The planet's diameter was known fairly accurately, but its mass could not be determined very accurately because there was no satellite circling Venus. The perpetual covering of clouds keeps the surface completely hidden. There was still speculation that the planet might consist of great swamps, be covered with oily wastes, or be a vast desert. The possibility that the dense carbon dioxide atmosphere would act like the glass of a greenhouse to entrap solar radiation had also been proposed by Carl Sagan and seemed a reasonable explanation for the high surface temperature indicated by the radio and radar measurements. Yet scientists were still not absolutely sure that the surface was so hot. The radar and radio data might be measuring the temperature high in the atmosphere. If so, the surface might be cool and relatively Earth-like.

Radar mapping of Venus from Earth contributed much information about the planet's surface which proved invaluable later in assessing radar images obtained by spacecraft. In 1970 a great swatch of Venus larger than the continent of Asia was mapped by radar astronomers of Caltech's Jet Propulsion Laboratory with the aid of the 64-meter (210-ft) dish antenna of the Goldstone Tracking Station in California's Mojave Desert.

The new radar map (figure 1.10) covered nearly half of the face of Venus visible during the Spring of 1970, extending from about 90° W to 30° E longitude, and from 45° S to about 35° N latitude. This area is about one-sixth of the total surface of the planet, which was mapped with about 100 times as much data as on the previous map made in 1968. The resolution was about twice as good as that which can be seen on the Moon with the naked eye.

To radar map Venus a radio signal at 12.5 cm wavelength was transmitted from Earth toward the planet.

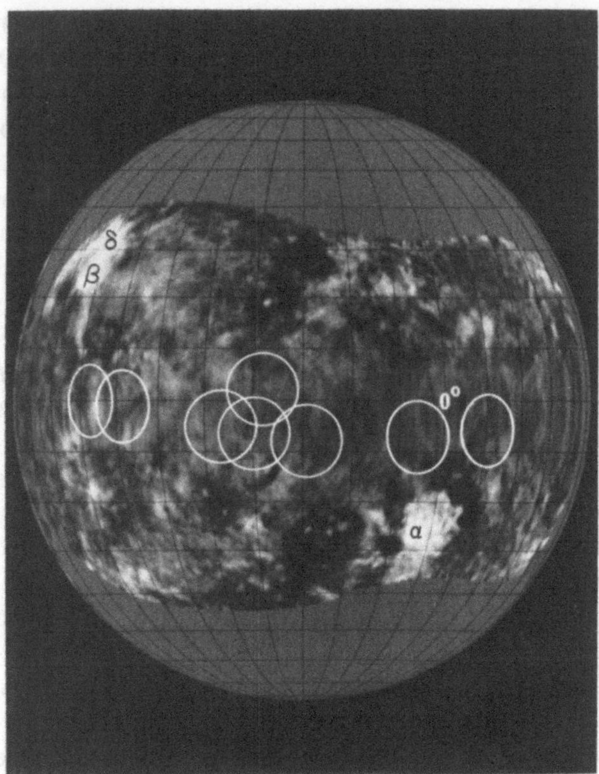

Figure 1.10. Very large radio telescope antennas on Earth were used to obtain the first radar images of the surface of Venus in the early 1970s. This picture, obtained with the antennas at Goldstone Tracking Station by R. M. Goldstein, is a composite map showing areas of Venus identified as Alpha, Beta, and Delta. The circles on the map show areas that were covered at higher resolution. (Photo NASA/JPL)

The large antenna allowed the energy in the radio waves to be focused into a tight "searchlight" beam. But by the time it reached Venus the beam was wide enough to envelop the whole planet. The radio signal bounced back off the surface of the planet, returning a mere whisper of an echo, a tiny fraction of a watt out of the 450 kw transmitted. When the signal returned unpolarized (i.e., the electric field was scrambled) a rough terrain was indicated. If unscrambled, the echo came from a relatively smooth surface.

Variations in time delay of the echo indicate changes in distance, and Doppler shift in frequency of the returned signals indicate motion toward or away from Earth. All the factors are combined to produce a two-dimensional representation of the mapping cells each about 5 km (3 mi) square on the surface of the distant planet. A computer program assembled all the data points into the radar map.

Prominent rugged areas seen on the map included Alpha, a region roughly 1600 km (1000 mi) in diameter located in the planet's southern hemisphere at about 25° S latitude and used to define the prime meridian. Besides Alpha there appeared to be as many as a dozen rough areas on the otherwise smooth surface of Venus. One of these, Beta, mapped in 1967–68, is about 320 km (200 mi) in diameter.

In June 1972, the next inferior conjunction, the radar probing continued. The resolution obtained was about 10 km (6 mi), five times better than the maps made during the 1970 conjunction. Huge shallow craters were discovered pockmarking the equatorial region of Venus. A 1500 km (930 mi) swatch of Venus, about the size of Alaska, showed a dozen craters up to 160 km (100 mi) across. The improved details of the surface were obtained by using two big antennas at Goldstone—the 64 meter (210 ft) dish and a 26 meter (85-ft) dish. The two antennas located 21 km (13 mi) apart permitted the radar scientists to use an interferometry mode which provided stereo reception and allowed them to determine elevation differences of 200 meters (650 ft). All the craters appeared shallow. The largest was estimated as 400 meters (1,300 ft) in depth. The area mapped showed a great valley on Venus (figure 1.11). The nearly horizontal dark swath on the high resolution image depicts a belt that cannot be accurately mapped from the location of the Goldstone Tracking Station.

Other important radar images of Venus were obtained in a continuing program at the National Astronomy and Ionosphere Center, Arecibo, Puerto Rico, using the 300 meter (1000 ft) radio telescope dish antenna which is suspended across a large bowl-shaped depression in the mountains. Figure 1.12 illustrates the type of resolution that was obtained by 1975. This area of Venus shows a large bright feature at about 65° N latitude, which was named Maxwell. It appeared very bright on the radar image to the right of a very dark region at the center of the picture. The bright areas were

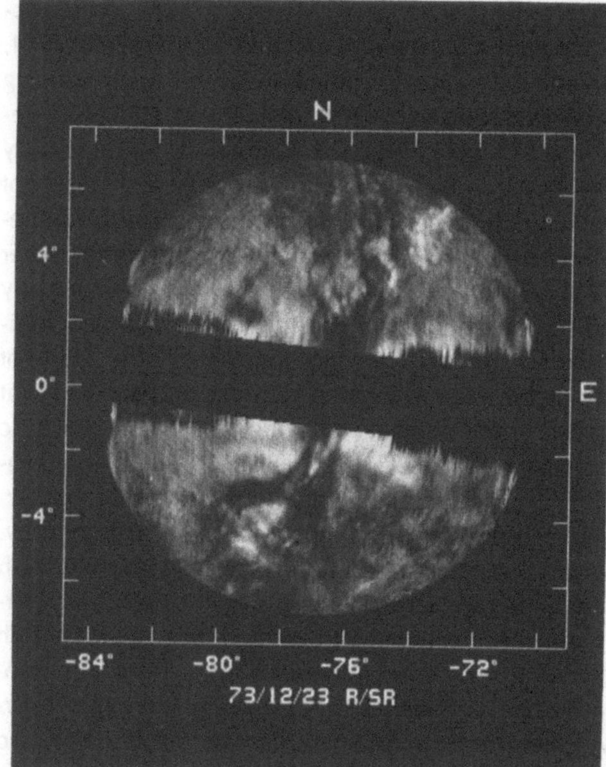

Figure 1.11. An area of Venus imaged by radar at high resolution shows a large valley stretching across the equator. Other high-resolution images revealed circular features that might be impact or volcanic craters. This image was obtained by R. M. Goldstein. (Photo NASA/JPL)

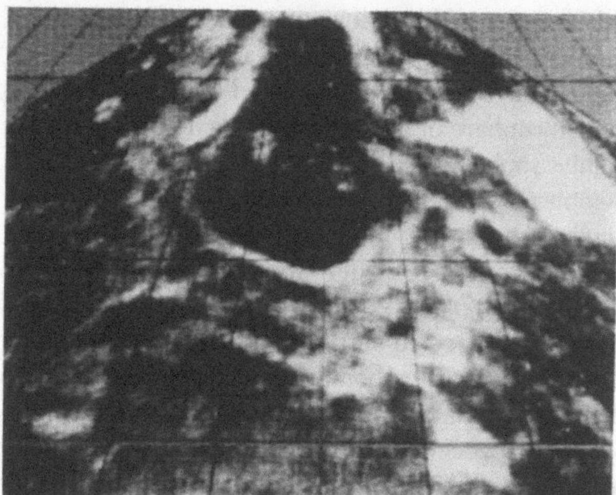

Figure 1.12. A Large radio relescope at Arecibo, Puerto Rico, has also been used to obtain radar images of Venus. This shows an area in mid-northern latitudes. The very bright feature on the top right was named Maxwell. (Photo NASA/ARC)

interpreted as being very rough and the dark areas smooth.

Since Venus is so Earthlike in size and probable evolution, scientists were particularly curious to find out why the planet appeared so different from Earth in its rotation, temperature, and atmosphere. The big question was whether or not Earth could follow the same evolutionary path, and what might cause such a change to Earth. Could such a change result from man's activities upon Earth, especially those which have resulted from the industrial revolution and the consequent more recent changes to Earth's environment and biosphere?

Answers could not be obtained from Earth. They could result only from taking scientific instruments to Venus, first in flybys of the planet and later by machines that would probe into the dense atmosphere and land upon the planet's surface.

At the beginning of the space age Venus was a prime target for planetary exploration. As was mentioned at the beginning of this chapter, the first, albeit scientifically unsuccessful, attempt was by the Soviets in 1961. First success came from an American spacecraft which zoomed past Venus in December 1962 and provided new and surprising data about the planet. In subsequent years the Soviets took nearly every opportunity to fly spacecraft by and to Venus while the American effort waxed and waned with infrequent visits to the planet. These various missions are described in the next three chapters, and the resulting new views of Venus are reviewed in later chapters. As a result of these spaceflights we now have a good idea of why Earth's errant twin followed the path it did to become the inhospitable world it is today.

After flyby spacecraft had provided sufficient new details to whet our appetites for more detailed information about Earth's intriguing sister planet, new technologies offered new capabilities for exploration. By the early 1970s the 20-year development of intercontinental ballistic missiles had provided the technological base for building scientific spacecraft capable of surviving the high temperatures and high deceleration forces that would be encountered by probes entering the atmosphere of Venus. This made it possible for the highly sophisticated instrumentation that had been demonstrated so successfully in other American spacecraft to be carried through the atmosphere of Venus down to its surface. A new approach to exploration of the shrouded planet could be taken. Thus the time was ripe for multifaceted missions to orbit Venus and probe through its dense carbon dioxide atmosphere down to the heated surface.

While spacecraft could not answer every question about Venus, they would take us closer to understanding the planet and why it differs from our Earth. Perhaps the most important aspect of this type of planetary exploration is that it provides us with details of extreme cases of conditions that in some ways resemble Earth. Venus (along with Mars) provides important comparisons with Earth. Part of the stumbling block to understanding our own planet has been our not knowing enough about other planets to make valid comparisons. The various space programs used spacecraft capable of carrying orbiters, probes, and landers in missions to Venus to help us make such comparisons.

In a presentation to the House Committee on Science and Astronautics on March 15, 1973 in connection with the NASA authorization for fiscal year 1974, when the U.S. Venus program was in danger of being canceled, Richard Goody of Harvard University repeated a statement he had made to the Royal Society in London on the occasion of the 500th anniversary of the birth of Copernicus: "It is no longer possible or desirable to consider Earth entirely aside from the other planets—planetary science has grown to contain many aspects of the earth sciences and for some geophysicists the aim of enquiry has now become the nature of the entire inner Solar System." He also stressed that some current attempts to model and predict climatic changes on Earth were stimulated directly by observations of planets such as Mars and Venus and that further in-depth study of Venus should have high priority.

Such exploration is important because while Earth and Venus are nearly twins physically, their characteristics today are extremely different. Earth is a water-rich planet on which life thrives. Venus is a dry and desolate world on which life as we know it could not survive, let alone thrive. Scientists want to know how two similar planets can evolve so differently and if there are any chances of Earth becoming like Venus. The question is surely also of great interest to the rest of humankind.

2

SPACE RACE

Because of the inherent difficulty of planetary missions, the U.S. National Aeronautics and Space Administration (NASA) assigned two launches of interplanetary spacecraft in 1962 in an attempt to achieve the first mission to another planet. The Soviets had failed in their 1961 attempt and there had been intense national political rivalry to gain space "firsts" ever since the Soviets launched the first artificial satellite of Earth in 1957.

Under contract to NASA, the Jet Propulsion Laboratory (JPL) of California Institute of Technology designed and constructed the Mariner series of interplanetary spacecraft capable of carrying science instruments to the terrestrial planets, Mercury, Venus, and Mars.

The Mariner A and B projects were intended to launch spacecraft to investigate interplanetary space and the vicinities of Venus and Mars respectively. The spacecraft were to be launched by a modified Atlas D intercontinental ballistic missile (ICBM) equipped with a Centaur high performance upper stage. Flights to Venus, as the first interplanetary objective, were scheduled for the summer of 1962. However, problems developed in delivering an operating Centaur upper stage on time for these launches. By the summer of 1961 it became apparent that the Venus mission would have to be delayed unless an alternative launch vehicle combination could be used. JPL suggested to NASA that an Atlas-Agena might be used instead of Atlas-Centaur if a less massive Venus spacecraft were developed from the Ranger spacecraft then being fabricated for lunar landings.

The energy required to leave Earth and travel to Venus is at a minimum approximately every 19 months. The position of Earth at launch for an optimum trajectory to rendezvous with Venus is approximately 60 degrees (sun-centered angle) ahead of Venus, i.e. prior to inferior conjunction. As launch dates and flight times deviate from the ideal, the energy required for a spacecraft to travel from Earth to Venus increases. This means that a smaller payload of scientific instruments has to be carried to Venus if launching is not at the optimum time.

NASA cancelled the Mariner A program and authorized the Mariner R project with the primary objective of launching two smaller spacecraft to Venus in 1962.

The technical and scientific objectives were to maintain two-way communication between Earth and the spacecraft on its journey to and beyond Venus, and to perform a scientific survey of the planet's gross characteristics during a flyby of the planet. Control of the spacecraft had to be precise enough to ensure that the unsterilized spacecraft would not crash onto the planet and contaminate its environment with Earth life forms. The Mariner R spacecraft weighed 204 kg (449 lb), much less than the Mariner A, and it was intended to carry about 11.3 kg (24.9 lb) of scientific instruments; later increased to 18 kg (40 lb).

Two spacecraft launches were scheduled to increase the probability of attaining a successful mission. The configurations of Earth and Venus on their orbits were such that the two spacecraft had to be launched within a period extending from July 18 through September 12, 1962. The time from the authorization of the Mariner R mission to the launch date was less than one year, requiring an all-out effort to design, develop, and procure components for the spacecraft and to test and launch the spacecraft within those few months. This was made possible by use of existing Mariner A and Ranger hardware and procedures. All major milestones were met on time and the ill-fated first spacecraft, Mariner 1, was launched on schedule on July 21. On August 27, the more fortunate Mariner 2 was successfully launched, and all project objectives were met with a high degree of success. The Mariner 2 spacecraft flew by Venus on December 14, 1962, and maintained radio contact with Earth until January 3, 1963.

The spacecraft was self-sufficient in power, using solar cells to convert enough solar radiation into electrical power to keep an internal battery charged. To communicate with Earth at the distance of Venus required a high-gain antenna, which had to be pointed toward Earth. The spacecraft used an attitude-control subsystem to orient itself in space by thrust from cold gas jets as needed. The spacecraft carried a small rocket engine which burned a storable rocket propellant. This rocket engine provided thrust for in-course maneuvers to correct the trajectory for the required flyby of Venus.

Because the spacecraft traveled toward the Sun, it required adequate thermal control through paint patterns, aluminum sheeting, thin gold plating, and polished aluminum surfaces to reflect or absorb the necessary amounts of heat. Aluminized plastic thermal shields protected the central compartment of the spacecraft. Polished aluminum louvers were automatically controlled to maintain sensitive electronic instruments at their required temperature.

The Mariner spacecraft was actually launched backward along Earth's orbit so that it fell inward toward the Sun, picking up speed and later passing Earth (figure 2.1). Venus, which travels faster than the Earth, overtook the Earth in October 1962 and caught up with Mariner in December 1962, for the encounter.

The engineers who designed the Mariner spacecraft had only theory on which to determine what conditions the spacecraft would face in deep interplanetary space. As satellites had explored the near space environment of Earth they had experienced many surprises. Would there be other surprises on the way to Venus? It was known that high- and low-energy radiation permeated space, but no one understood fully what effect this radiation would have on the spacecraft during its voyage. The effects of cosmic dust, micrometeorites, and larger objects, were unknown. Earth satellites had encountered collisions with micrometeorites. Closer to the Sun there might be more debris in space as suggested by the presence of the zodiacal light.

Another important question was the reliability of the spacecraft itself. Mariner was a new and complex machine that would have to work reliably millions of miles away from Earth without any possibility of maintenance or repair.

To obtain information for designing future spacecraft, Mariner 2 (figure 2.2) carried equipment to transmit 52 different engineering measurements made throughout the mission as well as science instruments to gather much needed data about the environment of interplanetary space toward the Sun. During the interplanetary cruise mode these instruments measured magnetic fields, cosmic dust, charged particles, and solar plasma. During the flyby of Venus, the engineering measurements were turned off and two primary scientific experiments sought information about the planet. These were radiometers to sense the heat coming from

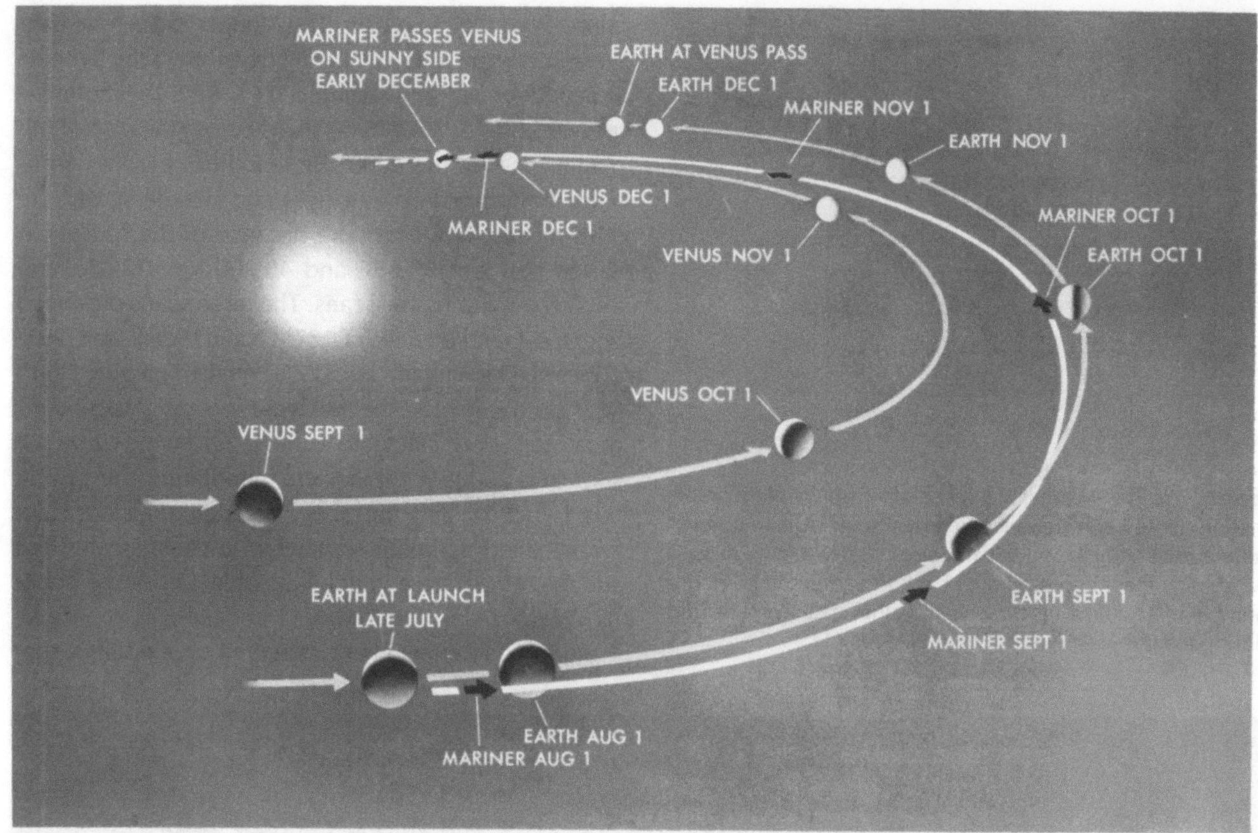

Figure 2.1. Trajectory of the first U.S. interplanetary spacecraft. Mariner 2 left Earth in August 1962 and reached Venus to fly past the planet in December of the same year. (Photo NASA/JPL)

the planet at infrared and microwave regions of the spectrum. The infrared radiometers covered bands centered on wavelengths of 10.4 and 8.4 micrometers. The microwave bands were at 13.3 and 19.0 mm. The cruise phase produced important new data about interplanetary space inside Earth's orbit. Only one impact of a micrometeorite was recorded, thereby demonstrating that interplanetary space inward from Earth is relatively free of cosmic dust. Mariner 2 verified that the solar plasma flows outward from the Sun as a stream of charged particles traveling at velocities varying from half a million to three million miles per hour. High-energy radiation in space was shown not to be dangerous. An astronaut could have survived the radiation dosage encountered by Mariner 2 on its voyage across the inner Solar System. The magnetic field in space did not vary much, and there was no appreciable change as the spacecraft approached Venus, indicating that the planet has a negligible magnetic field.

The encounter with Venus (figure 2.3) produced good data from the two radiometers as Mariner approached from 30 degrees above and behind the planet. Cloud temperature measured at the cloud tops was −50° C (−58° F) and about −34° C (−30° F) deep within the clouds. The difference in temperature between the light and dark hemispheres of the planet was negligible. On the basis of the radiometer scans, the surface of Venus, where the 19-mm radiation was believed to originate, appeared to have a temperature of about 127° C (260° F).

Microwave radiation from Venus observed earlier at Earth indicated a temperature of approximately 330° C (625° F), but infrared measurements made from Earth gave a temperature somewhat less than half that ob-

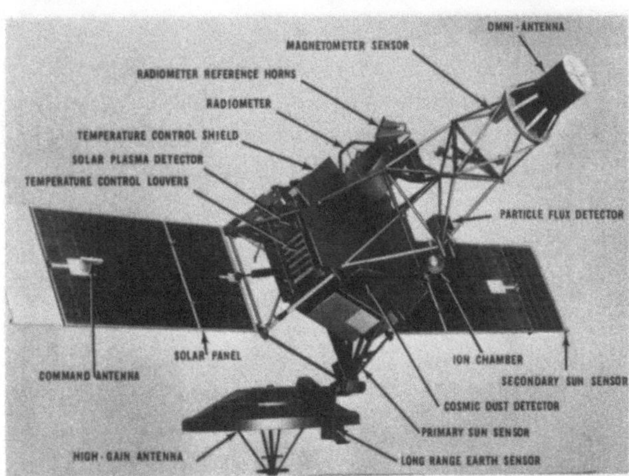

Figure 2.2. The Mariner 2 spacecraft, first to return information about Earth's errant twin, used solar cells to convert solar radiation into electrical power, carried a high-gain parabolic antenna to send data back to earth, and carried a number of scientific instruments to gather data in space and during the flyby of Venus. (Photo NASA/JPL)

tained from the radio data. The Mariner radiometer data were important to resolving the question of the temperature of Venus. On December 14, 1962, the radiometer was able to make three scans across the disc of the planet, one on the dark side, another near the terminator boundary between light and darkness, and the third on the light side. The distance of the spacecraft varied between 35,850 and 40,200 km (22,276 and 24,980 mi) during the scans. The three scans produced brightness temperatures of 187° C on the dark side, 297° C near the terminator, and 127° C on the light side (368°, 566°, and 260° F respectively). These brightness temperatures, were calculated on the assumption that the surface is 100 percent efficient in radiating energy. But this is not so and the above temperatures were acknowledged as being somewhat lower than the true surface temperature which was estimated to be about 425° C (767° F).

A limb-darkening effect, observed on the scans, sug-

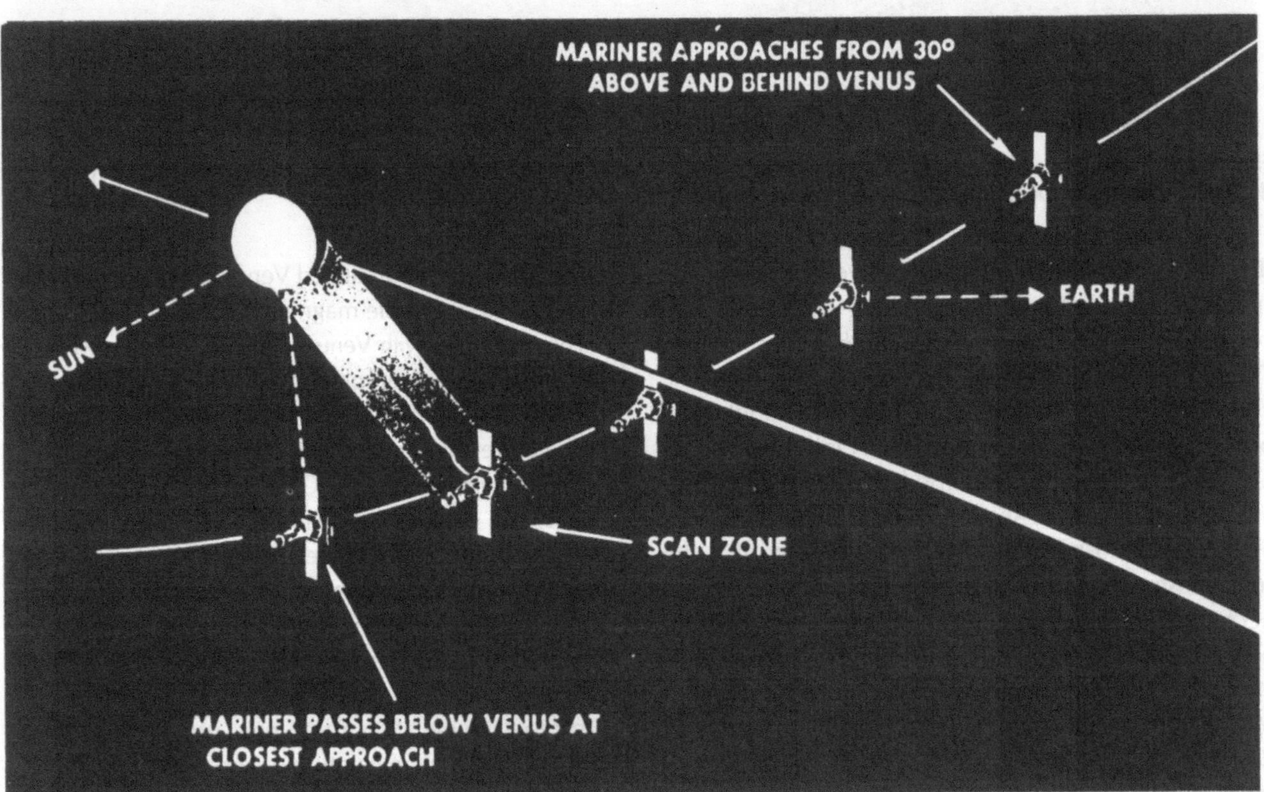

Figure 2.3. In its encounter with Venus, Mariner 2 approached from the dark side of the planet.

gested that high temperatures were deep within the atmosphere or at the surface of the planet. This is because when the radiometer looked straight down into the atmosphere there was the shortest path to the hot surface, whereas when it looked toward the edge (limb) of Venus it was looking through a thicker concentration of cool clouds. Thus the edge showed itself as cooler, or darker, to produce the observed limb darkening. This answered the question of whether the high temperatures of Venus as observed by radio from Earth originated at the surface or high in the atmosphere.

The flight of the spacecraft past Venus enabled scientists to make a more precise calculation of the mass of Venus as 0.81485 that of the Earth. Previously the mass had been calculated from the gravitational perturbation of other planets as 0.8148 that of the Earth.

During the 1962 mission Venus radar experiments continued from a station at Goldstone, California and radar signals were bounced off the planet. They showed that the surface of Venus is rough at wavelengths of 15 cm (6 in), and that the average dielectric constant of the surface material was similar to that of sand or dust, confirming experiments made the previous year. The radar bounce also confirmed the astronomical unit as 149,598,000 km plus or minus 480 km (92,956,000 mi, plus or minus 300 mi). The radar data also confirmed the slow rotation rate of the planet as 230 days plus or minus 40 to 50 days in a retrograde direction.

The success of the Venus mission had one disappointing result. NASA decided to modify its program of planetary exploration. Instead of repeating the Venus mission at the next opportunity in 1964, NASA decided to concentrate on other interplanetary projects, particularly the Mars mission of 1964, and send no more missions to Venus until 1967, when it was hoped a more advanced spacecraft would be available.

But the Soviets had no intention of neglecting Venus. They developed more advanced spacecraft for their next attempts to reach Venus. Zond 1 was launched on April 2, 1964. It attained a heliocentric orbit for a rendezvous with Venus, but it failed to return any planetary data. Western observers believed at the time that Zond 1 actually represented the eighth Soviet attempt to reach Venus, of which only two had been admitted by the Soviets. At any rate, Soviet engineers continued their efforts to improve the spacecraft, as they engaged in a spirited space race with the U.S. Soviet spacecraft were being conceived and designed for the exploration of the Moon and Mars as well as Venus (figure 2.4). Some of these engineering efforts were completed in time for the 1965 opportunity. A 963 kg (2123 lb) spacecraft, Venera 2, was launched on November 12, 1965, and the backup, Venera 3, which weighed slightly less, was launched on November 15, 1965. At the time, there was still great doubt about the composition of the sur-

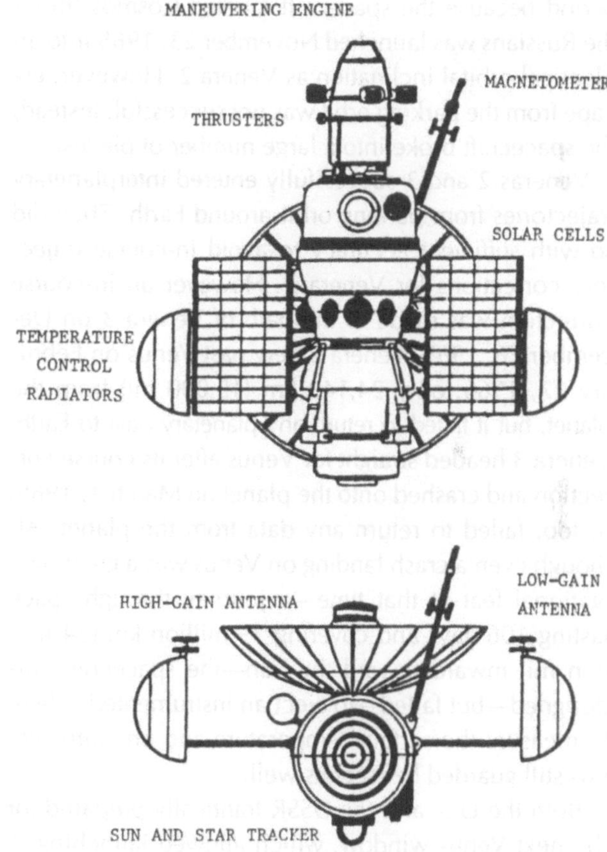

Figure 2.4. The Soviet spacecraft, Venera 2, was aimed to fly past the sunlit side of Venus and obtain photographs of the planet. Unfortunately communication failed and images were not returned. A similar spacecraft, Venera 3, struck the surface of Venus but, it too failed to return any data.

face, so the spacecraft carried surface phase sensors to determine if the capsule, when it landed on Venus, had landed on solid surface or in an ocean.

The aim of the new missions to Venus was to provide closeup radio studies of the planet, to determine the characteristics of the atmosphere and the surface, and to gather data on interplanetary space. The second spacecraft carried a parachute to achieve a soft landing. The lander had been sterilized before launch to prevent contamination of Venus by Earth organisms.

There was speculation at the time that a third Venus probe had been built and launched during this same period because the spacecraft labeled Cosmos 96 by the Russians was launched November 23, 1965 into an identical orbital inclination as Venera 2. However, escape from the parking orbit was not successful. Instead, the spacecraft broke into a large number of pieces.

Veneras 2 and 3 successfully entered interplanetary trajectories from parking orbit around Earth. They did so with sufficient accuracy to avoid in-course trajectory corrections for Venera 2. However an in-course correction was made to the path of Venera 3 on December 26, 1965. Venera 2 flew past Venus on February 27, 1966, only 24,140 km (15,000 mi) from the planet, but it failed to return any planetary data to Earth. Venera 3 headed straight for Venus after its course correction and crashed onto the planet on March 1, 1966. It, too, failed to return any data from the planet. Although even a crash landing on Venus was a great navigational feat at that time—a journey through space lasting 106 days and covering 39 million km (24 million mi) inward toward the Sun—the spacecraft was designed—but failed—to eject an instrumented sphere to measure the surface temperature and pressure. Venus still guarded her secrets well.

Both the U.S. and the USSR frantically prepared for the next Venus window, which allowed launching to the planet for a period of several months starting in June 1967. The Soviets had upgraded their endeavors to a highly ambitious project: they intended to land capsules on the hot surface of Venus and obtain information at the surface as well as during the descent through the Venus atmosphere. Venera 4 was the biggest interplanetary spacecraft yet. It weighed about 1086 kg (2,400 lb). Because early Venus spacecraft had lost communications, the Soviets approached the Director of the British Jodrell Bank Observatory, to ask for assistance from the Observatory's radiotelescope in monitoring transmission from the Venera as it plunged into the atmosphere of Venus. He agreed. The Russians were assured of support by one of the biggest terrestrial receiving antennas. The mission, launched on July 12, 1967, is described in detail in the next chapter, although some references are made to it below.

Meanwhile in the U.S., the Jet Propulsion Laboratory had designed and built a new Mariner (figure 2.5) based on experience with Mariner spacecraft that had flown successfully past Mars. While the Soviets were sending spacecraft to Venus at each available opportunity, a single spacecraft, Mariner 5, was all that NASA had allocated for the U.S. exploration of Venus for the decade following Mariner 2. Whereas the Mars Mariners had carried cameras to photograph the red planet, an

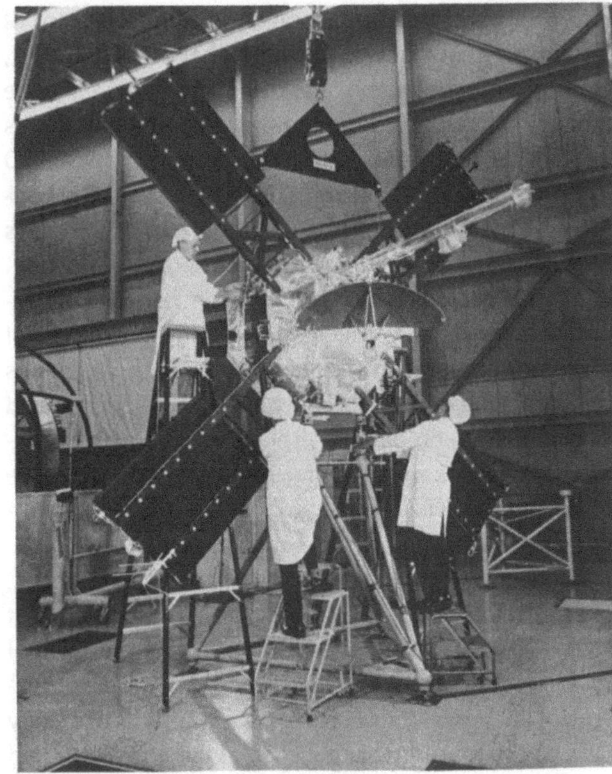

Figure 2.5. The Jet Propulsion Laboratory built a more advanced Mariner spacecraft for NASA. (Photo NASA/JPL)

unwise decision had been made about the Venus project: as Venus was shrouded with clouds, it was felt no cameras would be needed on Mariner 5. It turned out that this decision delayed for many years important discoveries about the circulation patterns of the clouds of Venus on which intricate patterns are observable in ultraviolet light.

Mariner-Venus 1967 was authorized as a project in December 1965. Equipment built to support Mariner-Mars 1964 (Mariner 4) was used so that maximum advantage could be taken of existing hardware and existing designs. Also, there would be only a single launch attempt. The primary objective of the mission was to fly by Venus and obtain scientific information that would complement and extend the results obtained by Mariner 2 in connection with seeking an understanding of the origin and of the environment of Venus. A secondary objective was to acquire engineering experience by converting a spacecraft originally designed for a flight to Mars into one to be flown to Venus and operating such a spacecraft on an interplanetary mission. The spacecraft was prepared for the Venus mission by converting the spare Mariner 4 spacecraft originally intended as a backup for a Mars flight. Seven science experiments were planned—an ultraviolet photometer, a solar plasma probe, a helium magnetometer, a trapped radiation detector, an S-band occultation radio, a dual frequency radio, and a celestial mechanics experiment based on tracking the spacecraft and observing its radio signals.

On July 14, two days after Venera 4, Mariner 5 lifted from Cape Canaveral and went directly into interplanetary orbit. Venera 4, by contrast, had entered its interplanetary trajectory from an Earth-parking orbit—the usual Soviet procedure.

On October 18, 1967, Venera 4 reached Venus ahead of Mariner 5. Its signals were received and recorded by Jodrell Bank's enormous radio telescope for an hour and twenty minutes. Then the probe entered the atmosphere of Venus and a shield of ionization generated by the enormous energy of entry blanketed the spacecraft. Its signals ended. During this blackout period a round instrumented package was ejected. Heavily insulated, it contained a braking canopy to slow its meteoric passage. An antenna on the top pointed back into space to transmit data over the 80 million km (50 million mi) to Earth.

Signals from the descent capsule were much weaker than those from the main spacecraft, but they were successfully received and carried data about the unearthly atmosphere of our sister planet. The entry and descent took place at the equator of Venus at longitude 80° E. The data showed that cloud temperatures of Venus ranged from 40° C (104° F) to 275° C (527° F). The surface pressure was at least 15 times that of Earth and the atmosphere consisted of 98.5 percent carbon dioxide. There was no trace of a magnetic field, nor of radiation belts such as those surrounding Earth. The exact level at which the Venera 4 probe stopped transmitting was not known, though at first the Soviets said that the probe had reached the surface.

On October 19, Mariner 5 flew 4023 km (2500 mi) above the surface of Venus at 1:34 P.M. EDT. As the spacecraft approached Venus its instruments (figure 2.6) began searching for the planet's magnetic field, measuring charged particles and identifying gases present in the upper atmosphere, and measuring levels of radiation. The approach velocity was relatively slow—only 3.05 km/sec (1.9 mi/sec). Combined with the close

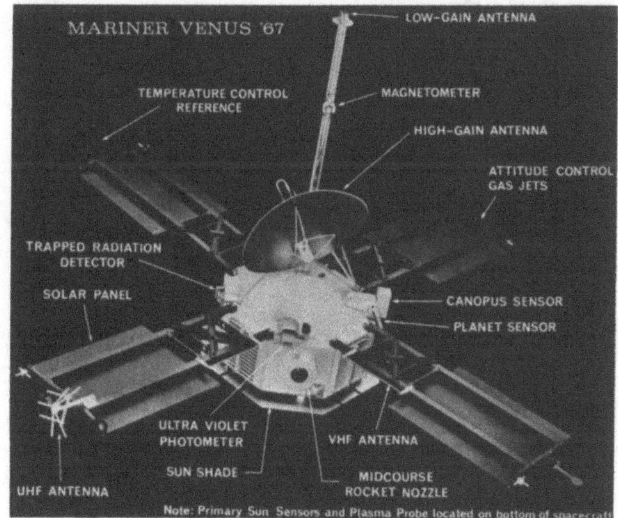

Figure 2.6. The new spacecraft, Mariner 5, carried a battery of science instruments to Venus in 1967. Unfortunately it did not include any camera system. (Photo NASA/JPL)

approach this slow speed led to a large deflection of the spacecraft's trajectory which, in turn, permitted a much more accurate determination of the mass of Venus, which was recalculated as 0.8149988 Earth's mass. Also it was found that Venus is much more nearly spherical than is Earth: Venus's oblateness was found to be 100 times less.

The encounter trajectory (figure 2.7) was from the dark side of Venus, ideally situated to find a magnetosphere of the planet, but no extensive magnetosphere was detected. However, the spacecraft discovered that the electrically conducting ionosphere envelope of Venus prohibits penetration by the magnetic field carried by the solar wind so that most of the solar wind plasma is prevented from reaching the atmosphere of Venus. A bow shock is formed in front of the planet and the solar wind is deflected around it. The solar wind does not penetrate to the surface of Venus as it does on the airless Moon.

As the flight path curved behind Venus cutting off signals, the radio signals penetrated deep into the atmosphere on their way to Earth (figure 2.8). This provided much information about that atmosphere, including the density, one of the prime measurement objectives of the mission.

In general Mariner's data confirmed that received from Venera 4. However, there were differences. The Mariner data suggested a lower percentage of carbon dioxide in the atmosphere of Venus. The spacecraft discovered a hydrogen corona on the dayside of the planet and recorded a weak magnetic field whereas the Soviets said Venus had no magnetic field.

Most importantly, calculated temperatures and pressures at the surface differed. From all available information the assumption had to be that the Venera 4 radioaltimeter had malfunctioned or its data were misinterpreted and the telemetry did not extend to the surface of Venus. It appeared that the Venera data ended some 26 km (16 mi) above the surface, and not at the surface. Thus the temperature and pressure at the surface were actually higher than those derived from the Venera probe data. There was also the possibility that

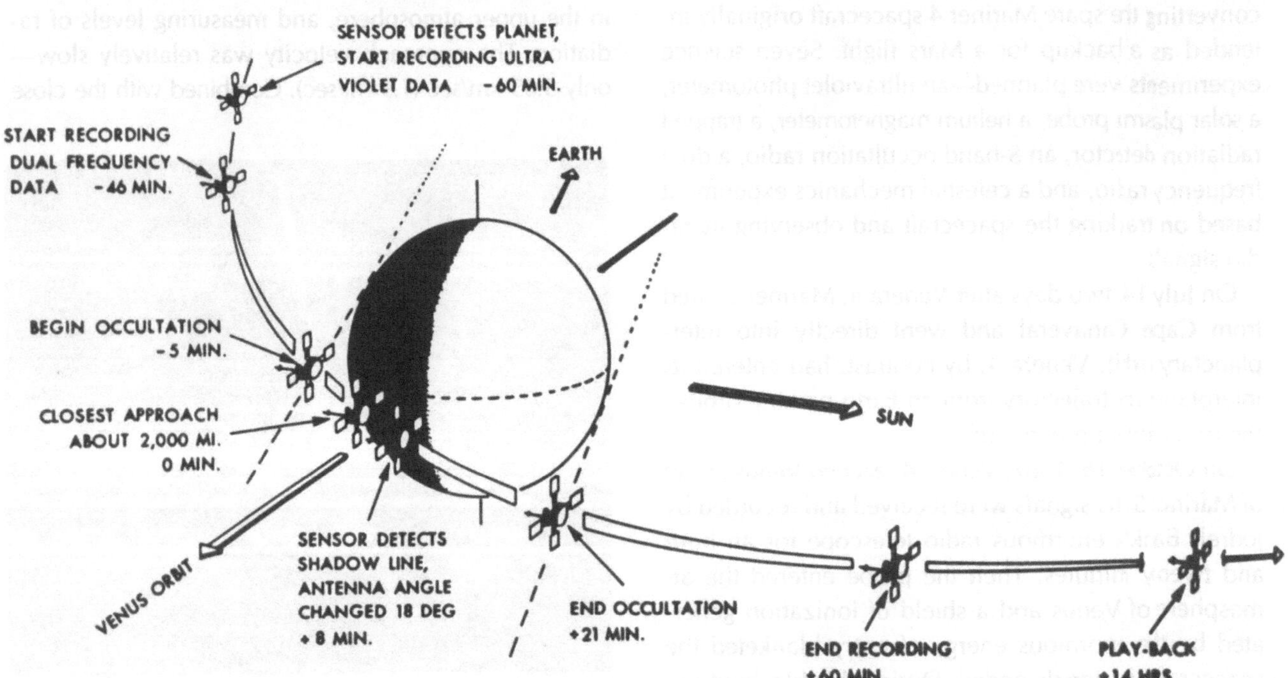

Figure 2.7. Again the Venus spacecraft approached the planet from the dark side. A radio occultation allowed scientists to gather data about the atmosphere as the spacecraft flew behind the planet as seen from Earth. (NASA)

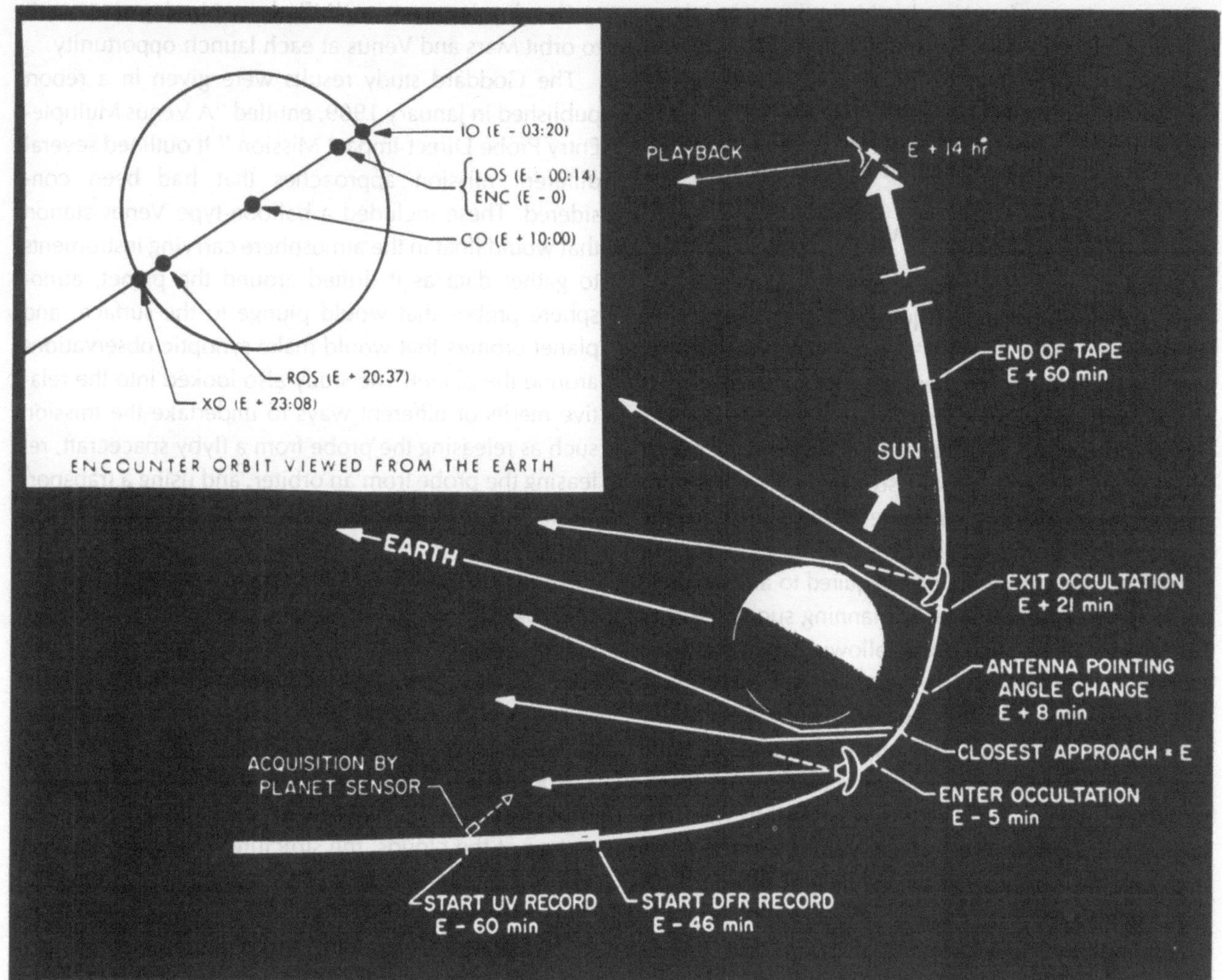

Figure 2.8. Times of occultation are shown in this drawing relative to the time of closest approch. (NASA)

the probe had landed on a 26 km (16 mi) high mountain. But radar mapping from Earth showed no unusual radar features in the area where the probe descended.

In summary the Mariner results produced several important facts about Venus.

- The solar wind flows around Venus without striking the planet's surface, contrary to what had been generally expected if the planet did not possess a magnetic field.
- The absence of energetic particles and radiation belts around Venus suggests that the magnetic field of the planet is less than 1 percent that of the Earth.
- There is an amount of atomic hydrogen in the upper atmosphere of Venus comparable with that in the upper atmosphere of Earth. However the hydrogen corona is much less extensive than that of Earth. There is no evidence of the presence of atomic oxygen.
- A weak ultraviolet airglow observed on the night side of Venus could be the source of the ashen light seen from Earth by earlier observers.
- Calculation of the mass of Venus was further refined to 0.8149988 that of the Earth. The equatorial radius of Venus does not differ by more than a few kilometers from 6053 km (3761 mi) which implies that the cloud tops are at an altitude of 67 km (42 mi).

- The percentage of carbon dioxide in the atmosphere is probably between 85 and 99 percent. Nitrogen cannot exceed a few percent, and oxygen and water vapor are even less abundant. The reason why the atmosphere is so opaque to infrared radiation is the presence of so much carbon dioxide. The surface temperature is between 380° and 530° C (716° and 986° F), and the surface pressure between 60 and 120 bars.

Nevertheless, significant questions about Venus remained unanswered. What are the composition and depth of the clouds? Are there other major constituents of the atmosphere and if so, what are they? Do any cool places exist on the surface, for example at the poles? Why are water and oxygen so scarce in the atmosphere? How does the atmosphere circulate about the planet? What is the nature of the surface? More advanced missions to Venus were required to answer these questions. Russia was already planning such missions, but NASA had no such plans following the success of Mariner 5. The Soviets, thwarted in their attempts to reach Mars successfully, seemed determined to explore Venus to the full. This appears strange in retrospect because Mars was a much easier planet to explore than Venus. Landing a spacecraft on Mars and having it operate on the surface was very much easier than landing a spacecraft on Venus and having it operate there.

Although there were no official plans for follow-up U.S. missions to Venus, a group of enthusiastic scientists started to study the feasibility of designing a simple entry probe that would be capable of penetrating the atmosphere of Venus and making measurements within that atmosphere. Centered at NASA Goddard Space Flight Center, this small group, under the leadership of Nelson W. Spencer, brought about the award of a study contract to AVCO Corporation. Also, a start was made to investigate the possibilities of using small Explorer type of spacecraft to explore the planets rather than the large Mariner class of vehicle.

By June 1968 a significant study had been published by the Space Science Board of the National Academy of Sciences. One of the recommendations was for NASA to develop inexpensive IMP-class spinning spacecraft to orbit Mars and Venus at each launch opportunity.

The Goddard study results were given in a report published in January 1969, entitled "A Venus Multiple-Entry Probe Direct-Impact Mission." It outlined several different mission approaches that had been considered. These included a balloon-type Venus station that would float in the atmosphere carrying instruments to gather data as it drifted around the planet, atmosphere probes that would plunge to the surface, and planet orbiters that would make synoptic observations around the planet. The study also looked into the relative merits of different ways to undertake the mission such as releasing the probe from a flyby spacecraft, releasing the probe from an orbiter, and using a transportation bus that would merely carry the probe to the planet.

The conclusion was that a direct mission with a bus that released several probes before entering the atmosphere was the most reliable. The complexity of balloon stations ruled them out for the near future, and the complexity of launching the probes from an orbiter also ruled that method out for the time being at least.

The report stated that seven entry probes were required to solve the major questions about Venus—the nature of the clouds, the structure, chemistry and motions of the planet's atmosphere, and the conditions on the surface.

Also in 1969, Goddard Space Flight Center awarded a further contract to AVCO Corporation to study a probe mission to Venus using the Thor-Delta launch vehicle. By the end of that year NASA Headquarters had developed a plan to have one spacecraft that could be used either as an orbiter or as a bus to carry probes to Venus. This became the universal bus concept.

Another report was developed by scientists of the Space Science Board and the Lunar and Planetary Missions Board of NASA, entitled "Venus— A Strategy for Exploration." This 1970 report recommended that exploration of Venus should be emphasized by NASA in the 1970s and 1980s and that a Delta-launched, spin-stabilized, Planetary Explorer spacecraft should be the main vehicle for initial missions of orbiters and probes and landers. The strategy recommended that no more than two launches should be attempted at each launch opportunity, and hybrid missions, in which a spacecraft carried both an orbiter and a probe, should be avoided because of their added complexity and cost. The aim was to keep the cost of each mission below $200 million, thereby making it economically feasible to plan a series of missions rather than one expensive and complex mission. Two multiprobe missions were suggested for the 1975 opportunity, two orbiters for 1976–77, and a landing for the 1978 opportunity. Unfortunately the concept of a series of missions became an impossible dream because successive administrations and congresses did not allocate sufficient funds for the dream to be realized. In fact, a very much reduced program was accomplished, and then only with considerable slippage of launch dates because of funding delays.

But while scientists in the U.S. were planning and studying and budgeting and trying to get the Administration and Congress to authorize new starts for Venus missions, the Soviets were pressing ahead, developing more complex spacecraft that would put them far ahead in the exploration of Venus.

In the meantime, U.S. exploration of Venus received an unexpected boost from another Mariner spacecraft which was being designed to travel to the innermost planet, Mercury. To do so it had to make use of gravity assist from Venus which required a close encounter with that planet. Most important, Mariner 10 would be carrying a camera to obtain images of Mercury's surface. So far no spacecraft had returned images of Venus. Now was a golden opportunity to do so.

The gravity assist technique had resulted from more than 20 years of speculation, research, and engineering development. It made possible a mission to Mercury (figure 2.9) that was economically acceptable. By its use a meaningful payload could be launched to Mercury by an Atlas-Centaur. The mission was approved by NASA in 1969 and a project office was established at the Jet Propulsion Laboratory in January 1970. By July 1971 a contract had been negotiated with the Boeing Company for the design and fabrication of the

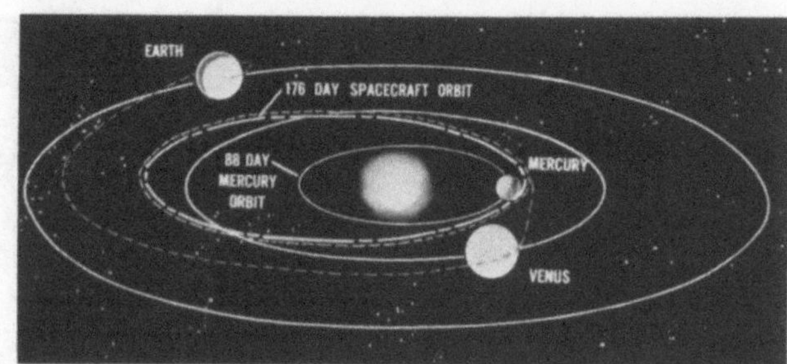

Figure 2.9. The mission to Mercury of Mariner 10 utilized the gravity assist obtained by a flyby of Venus. This gave American scientists an opportunity to explore Venus again despite the lack of American funds for a mission to Venus exclusively. Also the Mercury-bound spacecraft carried a camera system which presented an opportunity to obtain the first pictures of Venus from a spacecraft traveling near to the planet.

spacecraft. Again it was an economical single-spacecraft mission. The plan was to launch the spacecraft between October 16 and November 21, 1973, with a flyby of Venus between February 4 and 6, 1974, and an encounter with Mercury between March 27 and 31, 1974.

A launch window of about 90 minutes on November 3, 1973 would provide a mission with the best return of science data, and the 1973 opportunity offered one of the minimum launch energy requirements to swing by Venus and encounter Mercury. The television imaging system that had been used for the Mars flybys of earlier Mariners had to be redesigned because the flyby of Mercury would be on the night side of that planet. The spacecraft had also to be provided with better thermal protection so that it could function reliably within about 58 million km (36 million mi) of the Sun—the mean distance of Mercury from the Sun. A special Teflon-coated glass cloth was developed for the sunshade needed to protect the spacecraft.

At 12:45 A.M. EST on November 3, 1973, the Atlas-Centaur carrying the Mariner 10 spacecraft lifted into the night sky from the Florida launch site (figure 2.10). The Centaur separated successfully from the Atlas and its engines pushed the upper stage carrying the spacecraft into an orbit round the Earth. Partway along this

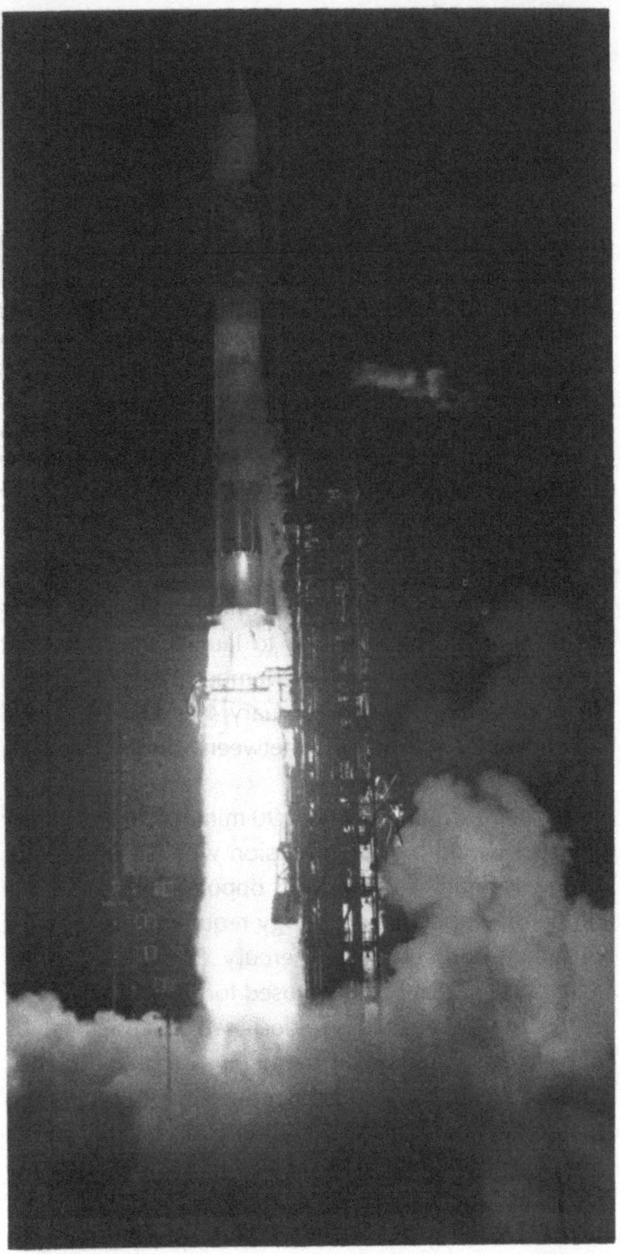

Figure 2.10. Mariner 10 was launched with an Atlas-Centaur on November 3, 1973. (Photo NASA)

orbit, the Centaur engine again fired and accelerated the spacecraft to the speed needed to break from the fetters of Earth's gravity and start its long voyage to Mercury. All seemed to be going well.

Then engineers became aware of a serious problem. Two thermal strap heaters surrounding the aluminum lens barrels of the TV cameras failed to turn on. Telemetry signals showed that the cameras were cooling down rapidly. Engineers and scientists were concerned that increasing coldness during the long voyage through space would damage the camera system or at least degrade its performance so that the pictures would not be as sharp as needed. Fortunately the temperature did not fall too much; it stabilized at what was considered to be an acceptable level. Nevertheless, to protect the cameras the scientists switched on the vidicons so that heat from them would help to keep the instrument warm until the spacecraft traveled closer to the Sun.

Tests were made of the optics system, and the cameras returned good pictures of Earth and Moon as the spacecraft left its home planet behind. Other instruments were also calibrated using Earth and Moon to make sure the instruments were operating correctly in space. Then the long cruise began with some of the instruments obtaining data of the environment of interplanetary space, the solar wind, magnetic fields, and charged particles.

By November 28 tracking showed that the spacecraft was heading too close to Venus for it to achieve gravitational swingby to Mercury. While an in-course maneuver was needed, the mission controllers were loathe to try such a maneuver. A week earlier they had commanded the spacecraft to roll on one axis so that instruments would be calibrated. During this maneuver the flight data system had unexpectedly reset itself to zero and the maneuver was canceled. The engineers tried to pinpoint the problem, and on Friday, December 7, 1973 another roll calibration maneuver was attempted. Again the flight data system reset. But this time the maneuver was completed and the high-gain antenna was calibrated by it. The in-flight maneuver was delayed for the time being.

A month later an even more serious problem developed. The spacecraft automatically and without warning changed from its main to its standby power system. This switchover could not be reversed by command from Earth. Should the backup power system then fail, the mission would have to be ended. Subsequently great care had to be taken in commanding any changes to the power status of the spacecraft to avoid stressing it and thereby increasing chances of a failure developing.

The high-gain antenna, essential to communicate images back to Earth, also suffered problems. Part of the feed system failed on December 25, 1973, and the signal power from the antenna dropped to an unacceptably low value. This would prevent transmission of images from Mercury as quickly as they could be obtained by the cameras—i.e., in real time—which would in turn mean that less of the planet's area could be photographed during the relatively brief period of the encounter. Engineers believed that low temperatures had caused the feed problem and from then on the spacecraft was oriented to warm the feed by solar radiation, with varying degrees of success.

On January 21, 1974, in response to a sequence of stored commands, Mariner 10 finally executed the in-flight maneuver needed to redirect its path. The correction ensured that Mariner would fly 5784 km (3594 mi) above the surface of Venus on February 5. This distance was the one calculated to bend the path of the spacecraft into a trajectory such that after the spacecraft passed Venus it would continue to the desired flyby of Mercury a few months later.

But the maneuver presented serious challenges. At one time oscillations built up in the gyroscope control system and the controllers were convinced that the spacecraft was dying. In the hour that it took to recognize and solve the problem, some 16 percent of the attitude-control gas carried by the spacecraft was discharged into space. This would considerably affect the maneuverability of the spacecraft and its ability to direct its cameras at the planets. Again the mission controllers worked quickly to develop alternative ways of controlling the spacecraft so that it would be able to meet all the science objectives of the mission.

The gyroscope problem had not been solved by the time the spacecraft had to be prepared for its encounter with Venus. At this time it was thought that the gyroscope had malfunctioned and that if the gyros were turned on again, the spacecraft would begin spinning uncontrollably. The controllers decided to attempt the flybv of Venus by the use of Sun and star references rather than the gyroscopes. But this, in turn, presented a major risk. Mariner might be distracted by the brilliance of Venus and swing around to lock on the light of the planet rather than the light from the star Canopus. Then the whole sequence of photography and scans of the clouds by other instruments would be upset and valuable information about Venus would be missed during the flyby. The risk had to be taken. Mariner sped toward Venus with its gyros idle and the starseeker locked on Canopus.

On January 17, 1974, the television camera heaters which had been inoperative since the start of the mission mysteriously turned themselves on. The unexpected "healing" was most welcome, because the now-jubilant imaging team members had feared that the cameras—with their temperatures dangerously low—would not operate for the encounter with Venus. Four days later the movable scan platform on which the cameras were mounted was calibrated on some stars. On February 4, with Mariner 640,000 km (398,000 mi) from Venus, the high-gain antenna feed problem suddenly righted itself also, but then soon afterward the signal strength again fell, although not to the same low level it had been earlier.

Despite all the problems encountered with the spacecraft, scientists were optimistic that the Venus encounter could be successful. The neurotic spacecraft had been nursed along, and controllers felt that they knew how to keep it functioning through the encounter. They were confident that they had invented ways to work around all the known problems.

On the day of the encounter with the Mistress of the Heavens—February 5, 1974—the planet rose slightly more than an hour before the Sun, depending on the latitude of an observer. It shone in the constellation Sagittarius as a bright morning star. Venus was nearly 45 million km (28 million mi) from Earth and appeared as a thin but growing crescent in a telescope, having passed through inferior conjunction and its close approach to Earth about the third week in January.

Mariner approached Venus from the dark side of the planet (figure 2.11). Just after midday EST the cameras of Mariner 10 starting taking pictures, but the images were of space. It was not until another half hour had elapsed that the first image of the planet was obtained.

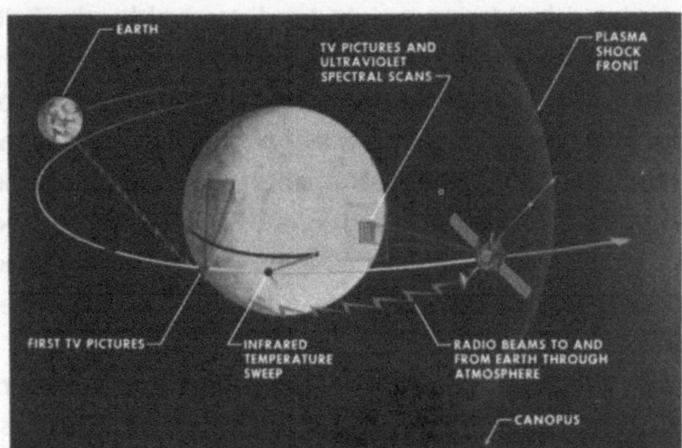

Figure 2.11. This spacecraft also approached Venus from the dark side of the planet so that the first pictures could be only of a very thin crescent. (Photo NASA/JPL)

The first picture was a high-resolution frame that showed only a small part of the planet, a fine cusp or horn of Venus close to the north pole (Figure 2.12). This image was obtained just before the spacecraft made its closest approach of 5790 km (3600 mi) to the planet. There were no protruberances or markings on this first picture of Venus from a spacecraft, nothing to indicate cumulus-type cloud tops or to show any cloud structure at the top of the planet's atmosphere.

Figure 2.12. The extremely thin crescent of Venus showed no detail in the clouds, and even though made at high magnification there were no cumulus type cloud tops in evidence. (Photo NASA/JPL)

The cameras continued to scan across the bright crescent of the planet to produce a sequence of long swathes that lay side by side and straddled the terminator and limb of the planet. They were for the most part featureless. The clouds of Venus appeared like planet-encompassing fog banks. However, those pictures, which showed the limb of the planet against the blackness of space, revealed interesting layers of haze at various levels above the top of the clouds (figure 2.13).

Six minutes after closest approach to Venus, the spacecraft went behind the planet as seen from Earth and its radio signals faded and then stopped, cut off by the bulk of Venus in the same way as for Mariner 5 earlier. The way in which the signals faded during this occultation was very important, because these radio waves passed through the atmosphere on their way to Earth. By studying the characteristics of the signal diminution with time at two radio frequencies, scientists obtained a completely new profile of Venus's atmosphere to great depths. Analysis of the radio signal indicated that the clouds had at least two different layers; a thin upper layer and a denser lower layer. Temperature inversions that might be associated with different cloud layers were also detected.

Figure 2.13. Above the ubiquitous cloud layer Mariner 10's cameras detected several layers of haze in the high atmosphere of the planet. (Photo NASA/JPL)

While behind Venus the spacecraft continued its picture-taking sequence, storing the images on magnetic tape within the spacecraft. After Mariner 10 emerged from behind Venus it transmitted the stored images to Earth, and when they were received they were a great surprise in the wealth of detail revealed. The earlier pictures taken by visible light before occultation had been most disappointing and ultraviolet pictures from Earth lacked good resolution. However, while on the sunward side of Venus, Mariner 10's cameras had continued its preplanned program and photographed the planet in ultraviolet light. The images being returned from tape storage and displayed on the monitor screens at the Jet Propulsion Laboratory revealed a whole new face to Venus. The ultraviolet images had a wealth of cloud detail; they showed intricate cloud patterns and polar hoods (figure 2.14). They exceeded the wildest expectations of some of the experimenters, who had previously had for study only indistinct faint patterns on photographs taken with the best telescopes on Earth.

Venus now became of great interest to meteorologists who were able to study these patterns and were presented with an opportunity to try to solve the question of atmospheric circulation. The patterns of Venus's clouds could now be compared with those of the cloud systems of Earth and Jupiter. The Mariner pictures confirmed the presence of C-, Y-, and Psi-shaped markings hinted at on the Earth-based pictures. They were now seen to be made up of much smaller detail (figure 2.15). At the location on Venus where the Sun shines directly down on the planet's surface, the ultraviolet images revealed polygonal cells which were attributed to rising air masses. Along the equatorial zone were fine streams of clouds. Both poles had hoods of clouds with spiral

Figure 2.14. After the Mariner spacecraft emerged from behind Venus, it transmitted to Earth pictures of the planet taken in ultraviolet light. They showed enormous details on Venus for the first time. (Photo NASA/JPL)

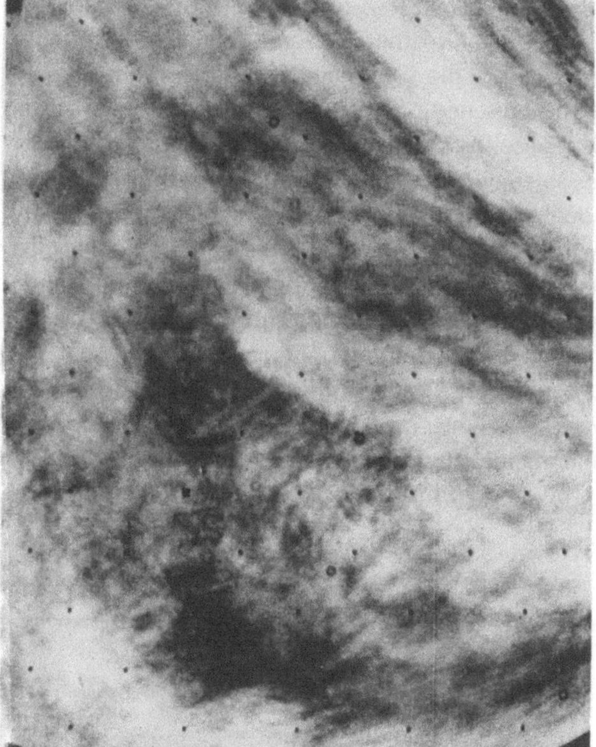

Figure 2.15. Close-up images revealed that the large markings seen indistinctly on images obtained by Earth-based telescopes actually consisted of a variety of intricate details. (Photo NASA/JPL)

patterns leading to them from lower latitudes. A series of pictures obtained over several days following the encounter confirmed the four-day rotation period of the clouds previously postulated from Earth-based observations (figure 2.16).

Having a camera in a spacecraft capable of returning images of Venus had certainly paid off enormously, thereby showing how erroneous had been earlier decisions not to carry cameras to the cloud-shrouded planet.

The other instruments carried by the spacecraft also returned important information about Venus. As Mariner sped toward Venus from the night side the instruments observed how Venus disturbs the magnetic field in interplanetary space and disturbs the flow of the solar wind streaming from the Sun. Venus forms a tail-like disturbance in the charged particles forming the solar wind. This tail stretches behind the planet away from the Sun and was recorded clearly by Mariner's instruments. Also the spacecraft found that the magnetic field of the solar wind is distorted considerably by the presence of Venus. The magnetic field of the planet itself is not, however, strong enough to deflect the solar wind as does Earth's magnetic field, even though the solar wind stream is modified considerably by the presence of Venus. Also, Mariner 10 confirmed that Venus does not have trapped particles in radiation belts like those surrounding Earth.

Again the scientists confirmed that somehow the ionosphere of Venus forms a bow shock on the sunward side of the planet which stops the solar wind from entering the atmosphere of the planet (figure 2.17).

The infrared instrument scanned from the dark side of the planet across the terminator to the sunlit side (figure 2.18) to measure the temperature of the clouds of Venus. This was approximately −23° C (−0° F) on both day and night hemispheres. Reduced infrared radiation near the edge of the visible disc confirmed the

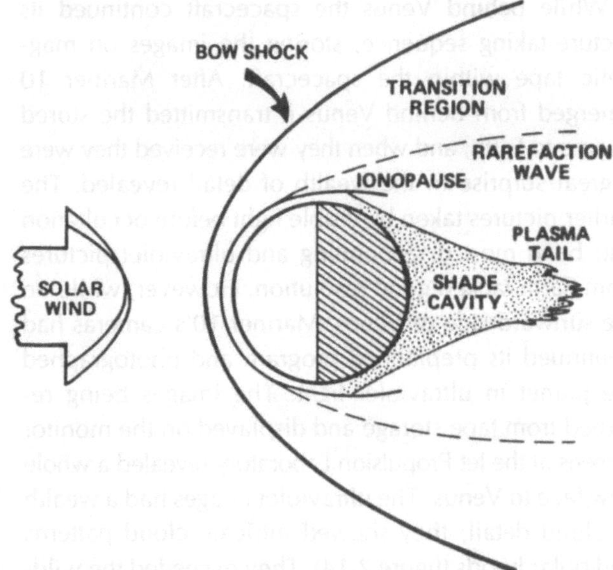

Figure 2.17. Venus was shown to have a bow shock and the solar wind was prevented from impinging directly on the atmosphere of the planet by an induced magnetic field.

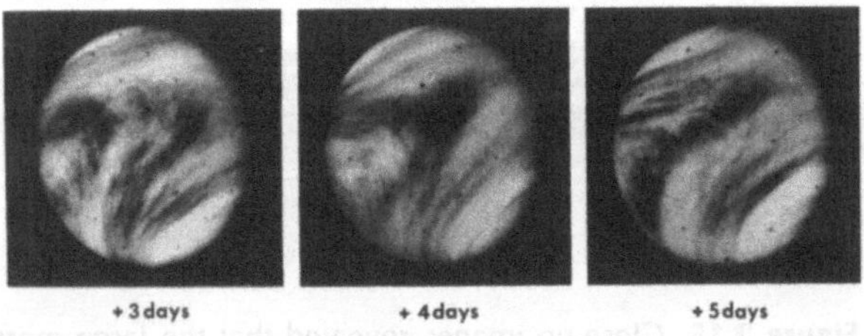

Figure 2.16. Mariner 10 obtained sufficient images to confirm the four-day rotation period of the ultraviolet features. (Photo NASA/JPL)

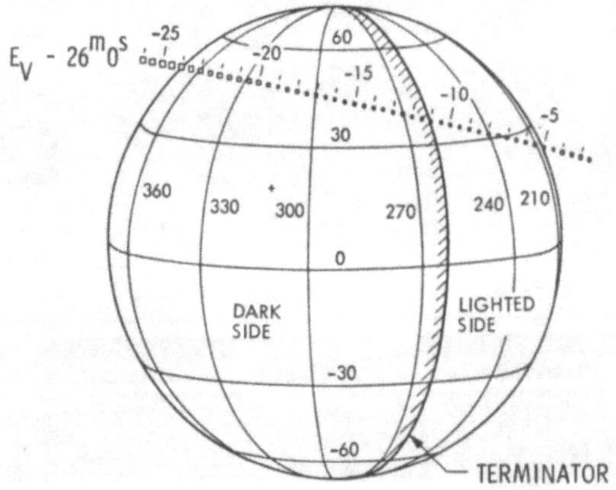

Figure 2.18. The infrared radiometer scanned across the terminator from the dark to the daylit side of Venus and revealed that the temperature did not vary greatly between day and night on the planet.

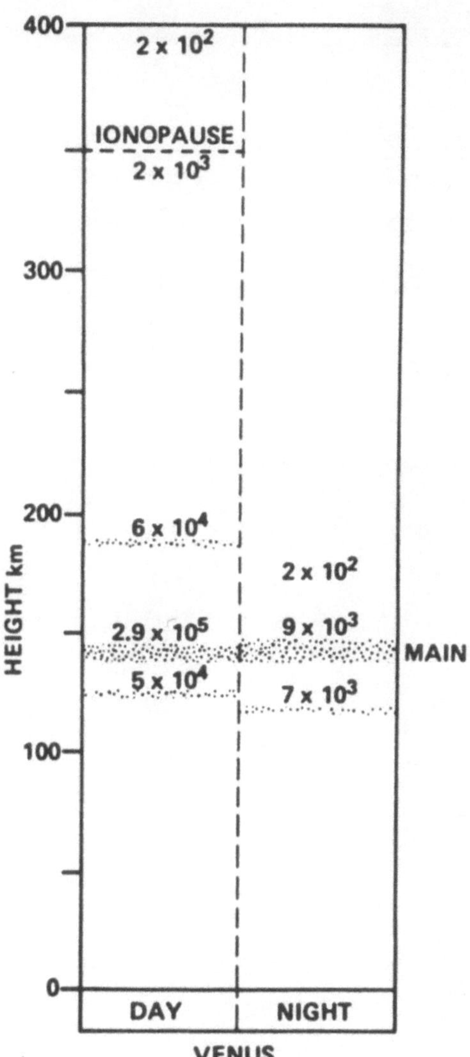

Figure 2.19. Mariner 10's flyby of Venus allowed scientists to determine that the ionosphere of Venus has peaks of ionization that differ in altitude on the night and day hemispheres.

limb darkening observed on earlier missions, showing that the atmosphere is very opaque to infrared radiation.

The ultraviolet airglow instrument sought evidence of important gases in the upper atmosphere. It found hydrogen, as anticipated, but no deuterium (heavy hydrogen), which implied that the hydrogen was derived from the incoming solar wind. Small quantities of helium were also detected and a relatively high concentration of atomic oxygen was present compared with Mars. The high oxygen concentration pointed to the upper atmosphere not mixing readily with the lower atmosphere. This stratification also seemed to be evidenced by the flatness of the cloud tops and the several horizontal layers above them.

Radio observations of the encounter confirmed the shape of Venus which was shown to be 100 times closer to a perfect sphere than is Earth. The occultation experiment revealed a lower cloud layer between 35 and 52 km (22 to 32 mi), and a quite different cloud layer with its top at 60 km (37 mi) above the hot surface of the planet. It also showed four regions of temperature inversion which appeared to be connected with the cloud layers.

The ionosphere of Venus (figure 2.19) was also investigated and shown to have a peak of ionization in the nighttime at 120 and 140 km (75 and 87 mi) and a stronger peak in the daytime at 145 km (90 mi).

After leaving Venus, Mariner 10 made three successful encounters with Mercury. But following Mariner 10, the American exploration of Venus had to rest for many years while the Soviets took over the active role with a series of highly sophisticated and successful orbiters and landers.

3

THE VENERAS: RED FLAGS ON VENUS

Since the beginning of the space age the Soviets have pursued an aggressive program to explore Venus. This exploration started in 1961 with the launching of Sputnik 7 on February 4, intended for a flyby trajectory. But the spacecraft failed to leave the Earth parking orbit. A companion spacecraft (Venera 1) launched February 12 successfully achieved an interplanetary orbit but, as described earlier, communication with the spacecraft ended within 25 days, when the spacecraft was about 23 million km (14 million mi) from Earth. Even the giant radio telescope at Jodrell Bank in Britain could not detect any signals from the spacecraft. Extrapolating the trajectory from the last known position the Soviets calculated that the spacecraft flew by Venus about May 20, 1961 at a distance of approximately 10,000 km (6,200 mi).

The Venera 1 spacecraft (figure 3.1) used solar cells and batteries for power, and its internal temperature was regulated by louvers. Its payload of science instruments included a magnetometer and charged particle telescopes. There were star and Sun seekers for navigation, and a high gain, parasol-type parabolic antenna designed to open on approach to Venus. Since it was the first interplanetary spacecraft, it carried an Earth globe of titanium alloy, and a Soviet emblem embossed in a metal plaque.

The next opportunity to launch a spacecraft to Venus was 19 months later, in August and September 1962. This was when the U.S. made its two initial launches. There were several Soviet launches in this period, but the Russians have never admitted that any of them were Venus probes. The suspected Soviet failures were from launches on August 25, September 1, and September 12, 1962.

Before the next launch opportunity, the Soviets tested at least one orbital launching platform, so that when the 1964 launch window opened the Soviets were ready with four Venus probes. The first lifted off on February 26, but the upper stage of the rocket booster malfunctioned. The next, launched on March 4, also had launcher problems. On March 27 the third try was labeled Kosmos 27. It was launched successfully and entered a suitable parking orbit for a Venus mission, but no interplanetary probe appears to have been released from this satellite.

The final launch of a spacecraft during the window took place on April 2. It was rewarded with success. The satellite orbit was attained and a spacecraft labeled Zond 3 (In Russian zond means probe*) was injected into a trajectory to Venus.

Communication with the spacecraft was good for almost two months. Suddenly, on May 14, 1964, the radio signals from Zond 1 stopped. Tracking data were extrapolated to indicate that the silent spacecraft passed within 100,000 km (62,000 mi) of Venus about the middle of July.

The failures were disheartening, but as part of a long-range plan to conquer space the Soviets prepared for the next launch opportunity which started in November 1965. The first launch was Venera 2 on November 12, followed quickly by Venera 3 on November 16. The first spacecraft was aimed to fly by the planet and radio photo-images back. Aside from the usual antennas and solar cell structures, it consisted basically of two payload compartments, one carrying batteries, radio communications and other electronics, and the other carrying a phototelevision system similar to those used successfully by the Russian scientists to get pictures of the Moon's surface. The second Venus spacecraft was designed to land on Venus. Its second payload compartment carried a heavily insulated descent sphere enclosing instruments to measure pressure and temperature, and some Soviet emblems to commemorate the first landing on another planet. Descent would be slowed by a parachute.

Both spacecraft successfully entered trajectories to Venus. Both made scientific observations of interplanetary space on the way to Venus. However, just before Venera 2 reached the target planet the telemetered data showed an unexpected and unexplained increase in internal temperature. Then communication failed. No information was obtained about Venus although the probe had returned much data about the interplanetary medium between Earth and Venus.

Meanwhile Venera 3 approached its target. A course correction on December 26 directed the spacecraft onto its impact trajectory. The plan was to detach the entry capsule just before the spacecraft entered the atmo-

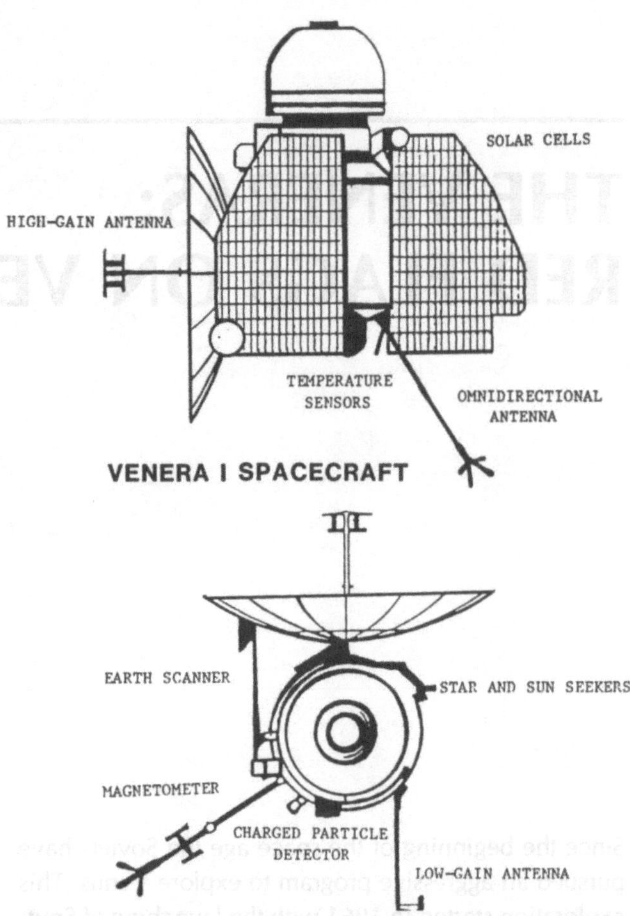

Figure 3.1. Diagram of the Soviet Venera 1 spacecraft which was launched in 1961 in an unsuccessful attempt to fly by the planet Venus. As with U.S. spacecraft, solar cells and a high-gain parabolic antenna are prominent features of the Venera.

*The term space probe is attributed to this author who first used it in an article entitled "The Martian Probe," published in the British aviation magazine *Aeronautics*, in May 1951. (See, NASA SP-4402 "Origins of NASA Names," 1976). The Mars probe article was available in Russia in 1957 in an official Russian translation of his book "Frontier to Space." For many years the term "space probe" was accepted for all interplanetary spacecraft. Recently, however, the term "probe" has been used more specifically in the U.S. to refer to spacecraft that penetrate planetary atmospheres.

sphere and was destroyed. But, as with Venera 2, communications problems plagued the spacecraft on the approach to Venus. The data link was lost. Again extrapolation of tracking data suggested that the spacecraft plunged into the atmosphere of Venus on the night side on March 1, 1966. Whether or not the lander capsule was ejected before entry is not known.

The Soviets may have planned a third mission for this launch opportunity because Kosmos 96 was launched on November 23, 1965 into an Earth orbit very similar to those used by the Veneras. It has been speculated that another entry probe was intended, but no spacecraft appeared to leave the satellite which itself broke up into a number of pieces. This suggested to Western observers that there had been a malfunction. The Soviets made no official comment.

During the following 19 months engineers and scientists worked hard to develop a more reliable and more sophisticated spacecraft design. The Soviets had also experienced some dismal failures in attempts to send spacecraft to Mars, despite an undoubted superiority in launch vehicles. A team led by Academician S.P. Korolev and later by G.N. Babakin made enormous strides forward in spacecraft design. The result was Venera 4, which was launched June 12, 1967. Again there appears to have been a double launch planned, because another spacecraft, Kosmos 167, was launched into a similar Earth orbit. While the Venera 4 probe was sent into an interplanetary trajectory to Venus, no probe appears to have left Kosmos 167, which reentered Earth's atmosphere after a few days.

Venera 4 (figure 3.2) was much heavier than its predecessors. The spacecraft weighed about 1100 kg (2425 lb) including its landing capsule. This capsule carried its own power supply to keep it operating for about an hour and a half after separation from the main spacecraft—sufficient time for it to reach the surface and continue for a short while to send data from that surface. Attempts had been made to avoid the temperature problems of earlier flights by forced gas cooling of the spacecraft. The usual interplanetary instruments were carried on Venera 4 for measurements of conditions in interplanetary space and in the vicinity of Venus. The landing capsule carried instruments to measure temperature, density, pressure, and the bulk composition of the atmosphere. A radio altimeter was included to determine the altitude of the landing capsule above the surface throughout the descent.

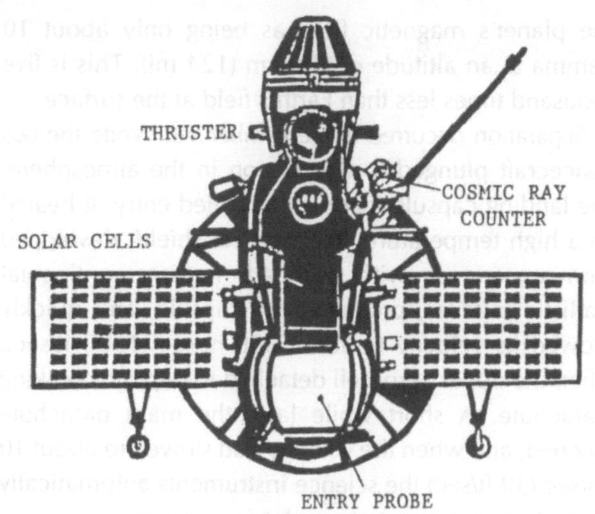

VENERA IV SPACECRAFT

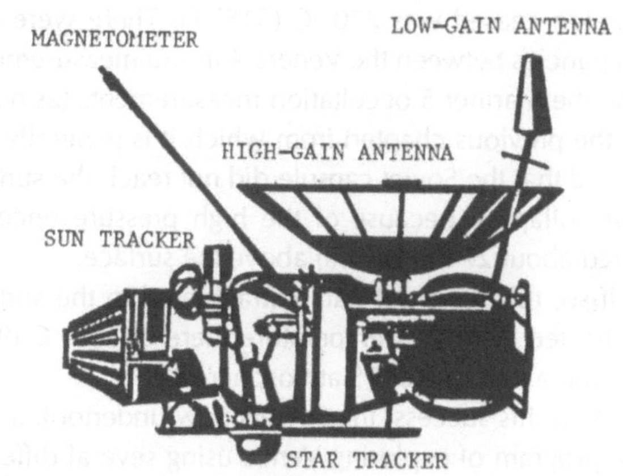

Figure 3.2. The Venera 4 spacecraft was more complex than its predecessors. The spherical object is a landing capsule that was designed to separate from the main spacecraft and plunge into Venus's atmosphere. Again solar cells and a parabolic high-gain antenna are prominent on the spacecraft.

On the approach to Venus, the spacecraft measured

the planet's magnetic field as being only about 10 gamma at an altitude of 200 km (124 mi). This is five thousand times less than Earth's field at the surface.

Separation occurred on September 18. While the bus spacecraft plunged to destruction in the atmosphere, the landing capsule made a controlled entry. It heated to a high temperature, but the heat shield, developed from experience with warheads for intercontinental ballistic missiles, protected the capsule, and quickly slowed its velocity to about 300 m/sec (1200 ft/sec). Then the upper aeroshell detached to deploy a braking parachute. A short while later the main parachute opened, and when the velocity had slowed to about 10 m/sec (30 ft/sec) the science instruments automatically turned on at a preset atmospheric pressure.

After the instruments were turned on, transmission of data continued for 93 minutes until it was assumed that the capsule had landed on the surface. A maximum atmospheric pressure of 18.5 kg/sq.cm (263 psi) was recorded, and the temperature at the time when transmission ceased was 270° C (518° F). There were discrepancies between the Venera 4 *in situ* measurements and the Mariner 5 occultation measurements (as noted in the previous chapter) from which it is generally believed that the Soviet capsule did not reach the surface but collapsed because of the high pressure encountered about 27 km (17 mi) above the surface.

If so, the Venera 4 data extrapolated to the surface indicated a surface temperature there of 500° C (932° F), and a pressure of 75 atmospheres.

With this success, the Soviets now undertook a major program of exploring Venus using several different types of spacecraft, orbiters, landers, and, soon to come, balloons. These were the same types of programs proposed for the U.S. by the NASA Goddard Space Flight Center study but which had not materialized because of on-again-off-again funding of American space programs. But the Soviet plans were on a much more ambitious basis apparently with less regard to the economic cost. While the U.S. Congress and successive administrations vacillated about missions to Venus and cut funding to the bone, the Soviets aggressively built a new breed of sophisticated spacecraft, some of which would conquer the searing heat of Venus and lift the veils from Earth's errant twin.

The Russian approach paid political dividends also. The Soviets were able to collect an imposing number of interplanetary firsts, as they had in unmanned exploration of the Moon and in manned orbital flight, which reflected an international image of technical prowess. Strangely, they were unsuccessful in missions to Mars and they do not appear to have attempted to develop a manned lunar program or missions to the outer planets at which the U.S. excelled, even though they had plans for or attempts to build huge boosters as big as the Saturn V.

The descent of Venera 4 through the atmosphere of Venus was a major breakthrough in planetary exploration and particularly in the exploration of Venus. Enormous technical problems had been overcome, among them that of how to protect a capsule from the vacuum of space and subsequently from intense pressures in Venus's atmosphere. This required the design of special seals.

Means to protect the entry capsule against the enormous heat of entry generated by interplanetary velocities was another major breakthrough. In addition, providing a direct radio link to Earth from the capsule plunging deep into the alien atmosphere represented a major advance in electronics communications. The Soviets certainly seemed to be pioneering these major aspects of technology. In retrospect it appears that the Soviets led the U.S. continually in space technology as opposed to space science, and continued to do so into the era of manned spaceflight, space stations, and plans to colonize other worlds.

It was a scant 15 months from the successful Venera 4 to the next launch opportunity. Soviet engineers and scientists mounted a crash program to develop the next generation of lander and orbiter spacecraft. There was insufficient time to do this for the 1969 launch opportunity, but development started for the opportunity after that—in August 1970. In the meantime improvements were made to the design of Venera 4, and two more spacecraft based on this design were readied for the January 1969 opportunity. This continuing program represented an enormous technical commitment on the part of the USSR—one far greater than any U.S. space commitment to planetary exploration. The Soviets benefited from a clearly defined long-range plan for expansion into space, whereas despite plans presented by NASA the U.S. efforts were thwarted by the whims of Congress and successive administrations with on-again-off-again type of interplanetary planning without long-range clearly defined goals.

The U.S. space program might be described as resulting from the dedication of small groups of scientists and engineers rather than from any imaginative government or political planning. And it was the unfailing dedication of these relatively few engineers and scientists which brought about the amazing U.S. accomplishments. Often the U.S. planetary explorations were made despite inadequate funding and political battles with the executive and legislative arms of government. Except for the manned expeditions to the Moon, pitifully few planetary explorations were acknowledged as having any priority by presidents or other political leaders. Even the historic first landing on Mars did not bring a U.S. President to the mission control center.

The triumphs of the American space program are monuments to these individual teams who made things happen on time and within costs. Had these men of vision been more fully supported by government their triumphs could have been greater, and the USSR would not have been the nation to achieve the first in many milestones of space exploration. Our grandchildren will probably ask why the U.S. was so slow in exploring space; and why after the political hysteria following the launching of Sputnik I, the first Earth satellite, every step was fraught with opposition from those who controlled the nation's purse strings. What had happened to the American spirit of private enterprise and exploration that had so successfully developed a whole continent in less than two centuries and could have gone on to develop the Solar System?

Undoubtedly the Russian scientists and engineers suffered greatly from the huge and unwieldy bureaucracy under which they worked, but they were assured of continuing support for many of their programs, especially those which fell within the objectives of the original plans for Soviet expansion into space and colonization of the Solar System. So while the American space scientists had to contend with small budgets and the need for much technical innovation to overcome shortage of funds, the Soviets were able to use large boosters and overkill to achieve their purpose. In the context of history we thus see many Soviet failures compared with the American space program. But the sheer magnitude of the effort inevitably gave many space firsts to the Soviets. The USSR used an expensive shotgun approach as opposed to the much less costly rifle approach of the U.S. The irony was that USSR scientists were provided with plenty of shotguns, whereas U.S. scientists were allocated only a pitifully few rifles.

As described in the next chapter, the success of Venera 4 generated criticism of the meager U.S. program to explore Venus. But instead of criticizing the paucity of the program the critics attacked the program itself, questioning the need to explore Venus since the Soviets were obviously being so successful. This negative attitude had to be fought aggressively by those U.S. scientists and engineers who wanted to keep the U.S. current in planetary science and technology.

The final configuration of Veneras 5 and 6 improved on Venera 4, but the main purpose of the spacecraft was to repeat the experiments of Venera 4 and consolidate the information obtained about Venus to that time. The transportation bus was almost the same as for Venera 4, with solar panels to keep chemical batteries charged, a high-gain parabolic antenna to be used when approaching Venus, and a sophisticated attitude-control system to orient the spacecraft, the solar panels, and the high-gain antenna.

Insulation on the 1-meter-diameter landing capsule was improved as was the capsule's resistance to the crushing pressure of the deep atmosphere of Venus. Venera 4 had failed at a pressure of about 20 atm., so the new capsules were designed to resist 27 atm. pressure and to penetrate the atmosphere more quickly. The Soviets it seems had still not accepted a surface pressure of 100 atm. An improvement was also made to the radar altimeter. A primary function was to gather data below 27 km (17 mi), a task at which the Soviets by this time conceded that Venera 4 had probably failed because of the high pressure. The two spacecraft were even heavier than their predecessors, each weighing about 1139 kg (2511 lb).

On January 5, 1969, Venera 5 entered orbit around Earth. After less than one orbit the interplanetary spacecraft separated from its orbital launch platform and was on its way to Venus. Five days later, Venera 6 left Earth and was injected into its interplanetary trajectory.

When Venera 5 was about 37,000 km (23,000 mi) from Venus, the descent capsule separated from the bus. It entered the atmosphere on May 16, 1969 on the dark side of the planet and rapidly decelerated. At a velocity of 210 m/sec (690 ft/sec) the drogue parachute deployed, to be followed shortly afterward by the main parachute. Then the radio altimeter antenna was extended and scientific instruments began to gather data on temperature, pressure, density, and composition, and to telemeter the data directly back to Earth. These data flowed back for about 53 minutes until, at a pressure of 27 atmospheres, the probe was crushed. But it had penetrated somewhat further than Venera 4, to an altitude of about 24 km (15 mi).

One day later Venera 6 repeated the performance of its companion, after separating from its bus somewhat closer to Venus than its predecessor. Its entry was also on the night side of the planet, and data were telemetered for 51 minutes until the capsule was presumed crushed at an altitude of about 10 km (6 mi) above the hot surface, but at virtually the same ambient pressure at which Venera 5 had failed. At first this suggested that Venera 6 had descended onto a mountainous area, but subsequently this was shown, following radar mapping of Venus, not to have been the case. Veneras 5 and 6 had probed toward the surface just south of the equator at longitudes 18 and 23° respectively, both of which localities are in the low-lying region north of Alpha Regio and east of the Aphrodite highland area. Venera 4 had probed toward another low-lying region at 19° N latitude and 38° longitude.

An important result from the science experiments of Veneras 5 and 6 was to confirm the percentages of atmospheric constituents. The gas analyzers carried by the probes measured concentrations of carbon dioxide, nitrogen, oxygen, and water vapor at two separate altitudes. The measurements showed a carbon dioxide concentration of about 96 percent, less than 5 percent for nitrogen and other inert gases, and less than 0.4 percent for oxygen. Extrapolated to the surface, the temperature and pressure measurements made by Venera 5 indicated a surface temperature of 530° C (986° F) and a pressure of 140 atmospheres; Venera 6 results were considerably different, suggesting that the altimeter had given spurious readings in showing that the spacecraft had penetrated to only 10 km (6 mi) above the surface and not to the surface itself.

So far the Soviet program had proved that Venus is a totally inhospitable planet to man. There could be no possibility of manned missions to the planet and it was unlikely that there were any forms of life on its surface. But while some might argue that the exploration of the planet should then have continued at a more leisurely pace, as was characteristic of the American planetary program, the Soviets continued to develop bigger and better spacecraft to explore Venus. Many questions about Venus were still unanswered, and foremost among them was why the planet, a veritable twin of Earth, had evolved so differently. This question obviously intrigued the Soviets and prompted them to continue their efforts to probe deep beneath the clouds of this alien world and find acceptable answers. But they were faced with the task of developing new types of space vehicles to do so.

While some of the problems about Venus had by this time been solved, a new set had arisen. Flights to Venus had demonstrated that it was possible to send probe spacecraft deep into the planet's atmosphere with a possibility of landing and surviving on the surface. Scientifically the probes had discovered that although Venus is similar to the Earth in physical parameters, climatic and atmospheric conditions are quite different from those of Earth. The challenge was to find the reasons for these differences and particularly to seek understanding of whether or not the climate and composition of Earth could change to become like those of Venus. If such a change were possible, what might cause it, and how might that cause be triggered? Would man-made environmental pollution serve as such a trigger? These questions were one of the main reasons, according to Soviet scientists, why they and many other scientists worldwide considered the exploration of Venus so important.

The Soviet scientists stated that Venus provided a natural cosmic laboratory in comparative planetology in which large-scale processes of a type which cannot be produced artificially in any laboratory on Earth can be studied. A planet's atmosphere is a complex system in which many interrelated processes are active. For example, its composition is governed by how the planet formed initially, and how it subsequently outgassed volatiles from its solid body. The composition is also affected by reactions between the atmosphere and rocks and any bodies of water on its surface, and by the structure of the high atmosphere from which gases can escape into space under certain circumstances or can be captured from the solar wind. Many of the processes depend upon the temperature of the atmosphere which, in turn, is governed by how much solar radiation is received and how much is trapped within the atmosphere and prevented from being returned to space. An important question about Venus was whether the hot surface was caused by a greenhouse effect trapping solar radiation or originated from heat generated within the planet.

The Soviet scientists stated that a full understanding of what takes place on Venus required sophisticated chemical analysis within the atmosphere, exact measurements of the altitudes at which atmospheric process are prevalent, the altitudes and spectral regions at which solar radiation is absorbed, and the nature of the clouds which blanket the planet. In addition information was needed about the composition of the surface; was it rocky, or dusty, mountainous or covered with vast plains, cratered with impacts from space or molded tectonically? Had there ever been water on Venus, and if so, what had caused it to virtually disappear?

Discovery of the presence of shock waves in the solar wind near Venus, and the absence of any significant magnetic field of the planet (as expected from its slow rotation rate) immediately raised questions as to the nature of the obstacle that retards the solar wind and forms the shock. Earth is protected by its strong magnetic field. What was it that held the solar wind off from Venus and prevented it from plunging directly into the atmosphere of the planet? One speculation was that the solar wind induced electric currents in the ionosphere of Venus which, in turn, produced a magnetic field able to hold off the solar wind. Thus Venus might have an induced magnetosphere rather than one based on an intrinsic magnetic field. More detailed measurements over a wide span of time were required. This called for instruments to be carried in orbit around Venus because flyby missions did not provide enough information over a long enough period of observation.

Designers of the next generation of Soviet spacecraft (figure 3.3) for exploring Venus added more weight to the landing capsule, which was almost completely redesigned to withstand 180 atmospheres so as to be able to operate at the surface. The Soviets had finally accepted very high pressures at the surface and had actually overreacted and overdesigned the spacecraft. The aim was to achieve a soft landing on the surface

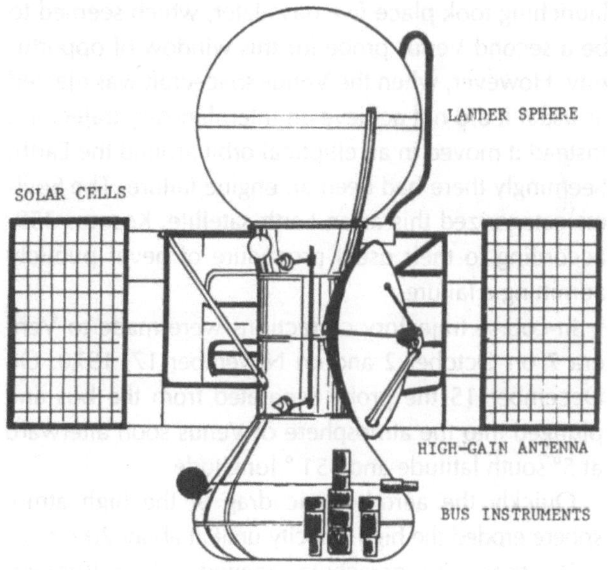

Figure 3.3. A new generation of spacecraft was developed in the early 1970s to explore Venus, following a number of failures of the earlier Soviet attempts.

and to gather data there. To do this the spherical landing capsule was equipped with shock absorbers. The main parachute was smaller than that used on earlier landers so that the capsule would descend more quickly and have a better chance to survive long enough to transmit a meaningful amount of data from the surface. Later this new generation of spacecraft designed to explore Venus would include orbiters to increase the synoptic coverage of the planet and its environment.

Venera 7 weighed 1180 kg (2600 lb) including its 500 kg (1100 lb) landing capsule. Beefed-up insulation was calculated to protect the capsule for a period of 90 minutes at 540° C (1000° F) before the internal temperature would become dangerously high. Also plans were made to reduce the temperature of the probe to below freezing before separating it from the bus to gain more operating time. Everything conceivable at that time had been done to try to ensure that the probe would reach and operate at the surface of Venus.

Venera 7 was launched successfully on August 17, 1970, and after reaching the parking orbit was quickly injected into the interplanetary trajectory. A second launching took place five days later, which seemed to be a second Venus probe for this window of opportunity. However, when the Venus spacecraft was ejected as usual it did not achieve an interplanetary trajectory. Instead it moved in an elliptical orbit around the Earth. Seemingly there had been an engine failure. The Soviets categorized this as an Earth satellite, Kosmos 359, according to their usual procedure of never publicly admitting a failure.

In-course trajectory corrections were made to Venera 7 on October 2 and on November 17, 1970. On December 15 the probe separated from the bus and plunged into the atmosphere of Venus soon afterward at 5° south latitude and 351 ° longitude.

Quickly the aerodynamic drag of the high atmosphere eroded the high velocity until at about 200 m/sec (650 ft/sec) the parachute opened. All instruments worked well, and volumes of important data flowed toward Earth. Unfortunately, when the capsule soft landed on the surface 35 minutes later, the data stream became garbled because the signal strength was one hundredth of that before impact. It seemed that the communication system had failed. However, very faint signals were detected and fortunately they were recorded for another half-hour. Later they produced some 25 minutes of valuable data after computer processing. An analysis of engineering data led to the conclusion that the landing had somehow affected the communication system or the antenna so as to reduce the power transmitted to Earth, probably because of the orientation of the spacecraft on the surface. The data received from the surface confirmed a surface temperature of 475° C (890° F) and a pressure of about 90 atmospheres.

At this time, too, the composition of the clouds of Venus was much in doubt and there were many speculations. Some claimed that the clouds were formed of ice crystals; others that the cloud particles were oxides of iron, mercury, or ammonia. Another view was that they consisted of dust particles whipped up from a hot desert of the planet by violent winds. Unfortunately these Soviet probes appear to have traveled too quickly through the clouds to gather significant information about them.

The Russians were justifiably proud of their accomplishment—one of many in 1970, which was a fantastic year for Soviet achievements: they had launched 85 vehicles into space compared with America's 29 civilian and military launches; there was only one manned American flight (the ill-fated Apollo 13 mission which failed to reach the Moon), whereas the Russians had dazzled the world with several manned orbital flights including a record-setting 18-day Earth orbiting flight in Soyuz 9, a prelude to Soviet continuous manned presence in space for over a decade.

Among unmanned space accomplishments were Luna 19's round trip to the Moon and the automatic return of a lunar soil sample, and placement of a robot vehicle to operate for many weeks exploring the Moon's surface.

At this time the United States had no missions planned to land spacecraft on Venus, and only sparse plans to explore Mars. But some scientists were thinking about trying to get a program started to send a probe to Venus and place an orbiter around the planet. JPL's Mariner 10, intended for a trip to Mercury, was seen as the sole opportunity for U.S. scientists to get instruments near to Venus in the next decade (as described in the previous chapter). Yet George M. Low, acting administrator of NASA boasted: "We still are the leaders in the exploration of space." And U.S. scientists were often quoted in the press during this period as saying that the Soviet craft could not have reached the surface of Venus, but would be crushed like the earlier abortive attempts at landing. Some claimed the Soviets would never be able to gather data from the hot surface.

But the main Soviet onslaught on Venus was still to come and would prove all the critics to be hopelessly unimaginative. New types of spacecraft were being developed in Russia that would completely surprise everyone by accomplishing on Venus what the Americans intended to accomplish on Mars—get landers to the surface, take pictures of the surface, and analyze samples of the rocks there. This would be a demonstration of a very advanced capability in space technology and remote control of robot vehicles over interplanetary distances.

But first there was Venera 8. This was prepared for the next launch opportunity and was very similar to Venera 7. The months before the launch opportunity had been used wisely. The experience with Venera 7 had been noted and used to upgrade the landing capsule further. Now that the temperature and pressure at the surface had been determined more precisely by Venera 7, the pressure overdesign could be reduced somewhat and additional heat insulation and science instruments could be carried. Whereas Venera 7 had been designed to withstand 180 atmospheres, Venera 8 was designed for 105 atmospheres only. Also to avoid communications problems after a landing, the capsule contained another antenna that would be ejected onto the surface on landing. Having two independent antennas would ensure that communications could continue with either antenna after the landing should a similar problem arise with the spacecraft orientation as with Venera 7.

New science instruments included a device to measure the light penetrating the atmosphere and on the surface to find out how much of the solar radiation was being absorbed by the clouds.

Another instrument would try to determine how much ammonia was present in the atmosphere. Horizontal wind speeds would also be measured, and the soil would be examined by observing its gamma ray spectra to check for radioactive elements.

The spacecraft was launched on March 27, 1972 and successfully entered an interplanetary trajectory. As usual a second launch followed on March 31, but as with the second launch of the previous window, interplanetary trajectory was not achieved—the spacecraft reached only an elliptical orbit around Earth. It was officially labeled Kosmos 482.

Venera 8 needed only a single trajectory correction on its way to Venus, thereby demonstrating again the ability of the Soviets to launch interplanetary vehicles accurately from Earth orbit. The descent capsule separated from the bus 53 minutes before encountering the atmosphere. The interior of the lander was cooled to −15° C (5° F). Together with the additional insulation of the probe this was expected to prolong operations on the surface. On July 22, 1972 the capsule encountered the upper atmosphere and started to decelerate. Target was 10° S latitude and 335° longitude on the dayside of the planet. Earlier probes had all entered on the night hemisphere. Again a quick descent was programmed so that the capsule could reach the surface before overheating. Landing took place in less than one hour after entry. Data poured back to Earth over the telemetry links.

Soviet scientists were jubilant at the great success of this mission. The engineers had conquered the problems that had plagued earlier attempts to land, and the way was opened for much more sophisticated scientific experiments on the surface of Earth's errant twin. The probe continued to send data from the surface for 50 minutes. Temperature at the landing site was measured as 470° C (878° F), very similar to the temperature on the dark hemisphere, and the pressure was 90 atmospheres. The new experiments also returned good data. Ammonia percentages of 0.1 and 0.01 percent at altitudes of 46 and 33 km (29 and 21 mi) respectively were recorded. Wind speeds had been measured throughout the descent and revealed winds of 100 m/sec (300 ft/sec) at altitudes above 48 km (30 mi), 40–70 m/sec (130–230 ft/sec) at altitudes between 48 and 42 km (30 and 26 mi), and a mere 1 m/sec (3 ft/sec) or less close to the surface. A large shear zone was thus recognized at what was thought to be close to the base of the cloud region.

A major discovery was that the surface of Venus is well illuminated despite the heavy cloud layers. The illumination fell rapidly in intensity at between 35 and 30 km (22 and 18 mi) but remained sufficiently intense at the surface to resemble that on a very heavily overcast day on Earth (figure 3.4).

With all this new information the Soviets were ready for their in-depth exploration of Venus. To implement the ambitious plans, they would have to miss a launch opportunity for the first time since they had started sending spacecraft to our neighbor world. In the ensuing months after the highly successful mission of Venera 8, Soviet engineers and scientists were hard at work perfecting a large and sophisticated Venus lander and an orbiter in time for the June 1975 launch window. The results of these expeditions would again confound the smug skeptics of the noncommunist world who should have known better and were, in fact, retarding

Figure 3.4. Illumination at the surface of Venus as measured by a Venera landing craft showed that sunlight penetrates to the surface. This artist's concept of a view on the surface of Venus, however, shows the sun, whereas it would probably appear only as a slight brightening of a region of the sky. (Photo NASA/JPL)

the progress of free enterprise science and technology. This author was reminded of a time during World War II when intelligence "experts" were stating that the Germans could not have developed a long-range rocket because they did not have the technical knowhow—and this only a few months before the V-2s began to fall on Antwerp and London. The lesson from history is that the technical capability of any advanced nation should never be underestimated. The question is always one of whether or not a national commitment is to be made and then supported. Advanced technology is never an exclusive possession of any society. Its denial to others forces them to develop their own capabilities as has been demonstrated in the past with nuclear capabilities and more recently in the high technology of digital computers and the software to operate them.

The next Venera spacecrafts were much larger. Four times the weight of Venera 8, the first of the new spacecraft weighed 4936 kg (10,880 lb), the largest interplanetary spacecraft ever. Its companion for this launch window was slightly more massive: 5033 kg (11,096 lb). These huge spacecraft required one of the largest launch rockets—the Soviet Proton booster, which had been used for many manned flights and was highly reliable. Instead of the major spacecraft being intended to fly by Venus or plunge to destruction in the atmosphere, it was designed as an orbiter. The lander which it carried to Venus was considerably improved by extensive redesign. It consisted of the usual sphere, now 2.4 m (7 ft 10 in) in diameter, surmounted by a parachute canister and an aerodynamic braking shield (figure 3.5). At the base of the sphere was a circular ring-type shock absorber system to reduce the effects of impact on the surface of the planet. The aerodynamic braking structure also acted as an antenna dish to telemeter engineering and science data to an orbital relay.

Many more instruments were now packed into the big lander. Most important was an imaging system to obtain pictures of the surface surrounding the lander, illuminated by floodlights carried aboard to alleviate the gloom expected on the basis of the measurements made by Venera 8. A densitometer was to be put onto

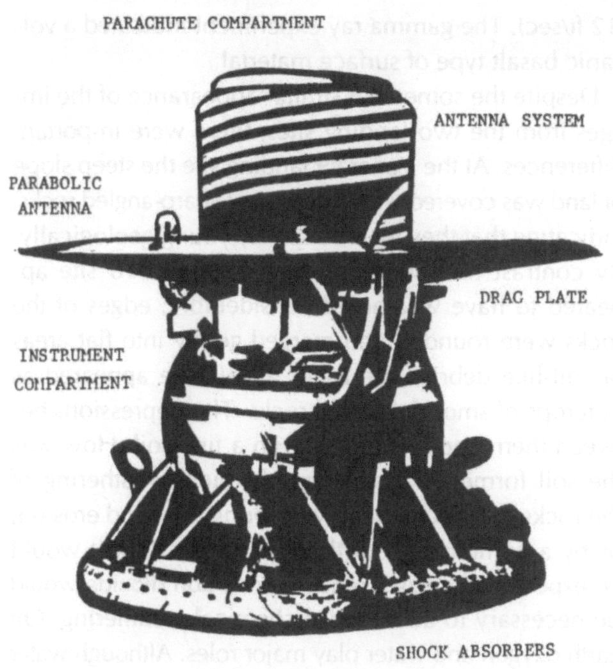

Figure 3.5. The Venera 9 spacecraft was the first to land safely on the planet and return pictures of the hot surface of Venus. This was a truly remarkable achievement which demonstrated a high technology.

the surface after landing to measure the density of the soil of Venus. The lander also carried a gamma-ray spectrometer, mass spectrometers, temperature and pressure sensors and a wind measuring instrument.

The orbiter carried a battery of science instruments to survey the planet from orbit. Cameras photographed the cloud tops in ultraviolet light to investigate the dynamics of atmospheric circulation. A photopolarimeter measured cloud brightness and polarization to determine cloud particle sizes. Infrared spectrometers detected heat radiation from the clouds at different parts of the spectrum to measure temperature and reflectivity. As in past flights a magnetometer and charged particle detectors were carried for use on the interplanetary cruise and to monitor the radiation and magnetic environment of Venus. The orbiter also acted as a radio relay for the transmissions from the lander.

June 8, 1975 the mighty Proton lofted the Venera 9 into Earth orbit from which the Venus-bound spacecraft was injected into an interplanetary trajectory. On June 12, a second huge Proton lofted Venera 10 into orbit, and that spacecraft, too, hurtled toward Venus. On October 20 the lander capsule of Venera 9 separated from the orbiter spacecraft, its interior already cooled in anticipation of the heat input from Venus. A new innovation was to also cool the exterior of the lander to − 100° C (− 148° F). The lander entered the Venus atmosphere on October 22, and the first braking parachute was ejected at an altitude of 64 km (40 mi). Later the main parachute opened and was subsequently jettisoned when the probe had penetrated the cloud layers. From then on the probe fell freely through the thick atmosphere slowed only by its airbrake until, 75 minutes later, landing was made safely on the dayside of the planet, at latitude 32° N and longitude 291°.

During each lander's descent through the atmosphere, its nephelometer provided information about the cloud layers of Venus. The experiment confirmed the altitude of the lower boundary of the clouds and provided good estimates of the sizes and concentrations of cloud particles. The clouds of Venus proved to be relatively transparent, more like mists than the typical clouds of Earth. But their layers are extremely thick. The upper layer extends from 70 to 57 km (44 to 36 mi) above the surface, the middle layer from 57 to 52 km (36 to 33 mi), and the lower layer from 52 to 49 km (33 to 31 mi). Particles were found to be of three types: ubiquitous large particles some 7 micrometers in diameter which accounted for about 90 percent of the mass of the clouds; medium-sized particles of 2 to 2.5 micrometers size; and small particles with an average diameter of 0.4 micrometers. While the large particles were present in all three layers, the medium and small particles were confined to the two upper layers.

The two landers also made important observations of the winds of Venus. Ultraviolet observations had revealed much earlier that the clouds move, presumably together with the upper part of the atmosphere, to rotate completely around the planet in a period of about 4 days. This meant that in the high atmosphere there were high velocities relative to the surface. The high wind speeds continued through the clouds, with a large wind shear at the base of the clouds, followed by a gradual decline in wind speed to about 1 m/sec at the surface.

For 53 minutes Venera 9 transmitted data from the surface of Venus and, most astonishingly, it took and transmitted the first picture of the alien landscape (figure 3.6). Diffused solar illumination was so bright at the surface that the floodlights were not needed to obtain the photograph. The Sun was 54 degrees above the local horizon at the landing site; by contrast, it had been only a few degrees above the horizon when Venera 8 gathered its data at the surface. The picture revealed a rock-strewn area with many sharp-edged flat rocks extending to what seemed to be a close horizon. The area was a volcanic one, on the slopes of the feature identified by Earth-based radar as Beta Regio which on the basis of today's knowledge appears to be a group of shield volcanoes.

The second lander, Venera 10, also made a perfect entry and safe descent through the atmosphere, landing on October 23, 1974 at latitude 16° N and longitude 291°, also in the Beta Regio area. It transmitted data for a slightly longer period—65 minutes—including another picture (figure 3.7) of the surrounding surface, again showing a rock strewn landscape but with a different type of rock from that at the landing site of Venera 9. Greater areas of debris were visible between the individual rocks as though more erosion had taken place because the surface was older.

At both sites wind velocities were less than 4 m/sec (12 ft/sec). The gamma ray experiment indicated a volcanic basalt type of surface material.

Despite the somewhat similar appearance of the images from the two landing sites, there were important differences. At the Venera 9 landing site the steep slope of land was covered with a deposit of sharp-angled rocks, indicating that they were relatively young geologically. By contrast the landscape at the Venera 10 site appeared to have weathered considerably; edges of the rocks were rounded and merged gently into flat areas of soil-like debris. The rocks at this site appeared as outcrops of smooth layered rocks. The depressions between them appeared filled with a fine soil. How was the soil formed? Possibly by chemical weathering of the rocks, by mechanical weathering, by wind erosion, or by a combination of these processes. But it would be expected that gases other than carbon dioxide would be necessary to account for chemical weathering. On Earth oxygen and water play major roles. Although water vapor had been detected by the probes, it was almost certain that there cannot have been liquid water in quantities to weather the rocks, at least for a time measured in millions if not billions of years.

Radioactivity measurements made by the landers indicated that the rocks had a high basaltic content. Such rocks could be weathered by water vapor alone to form hydrated minerals. The inert nature of carbon dioxide would not permit that gas, in the absence of much water, to accumulate in the form of limestone and other carbonate rocks as carbon dioxide does on Earth's surface.

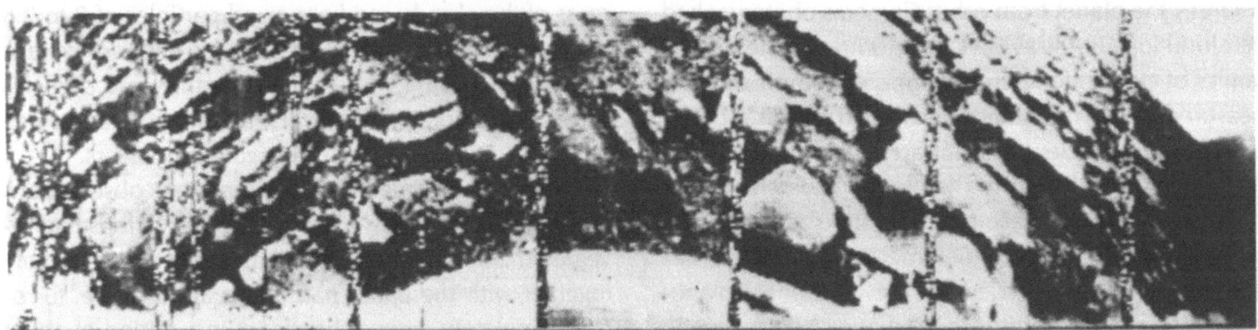

Figure 3.6. Panorama of the surface of Venus taken by Venera 9 showing a landscape of rocks with soil between them. The horizon and sky appear at the right top corner of the picture.

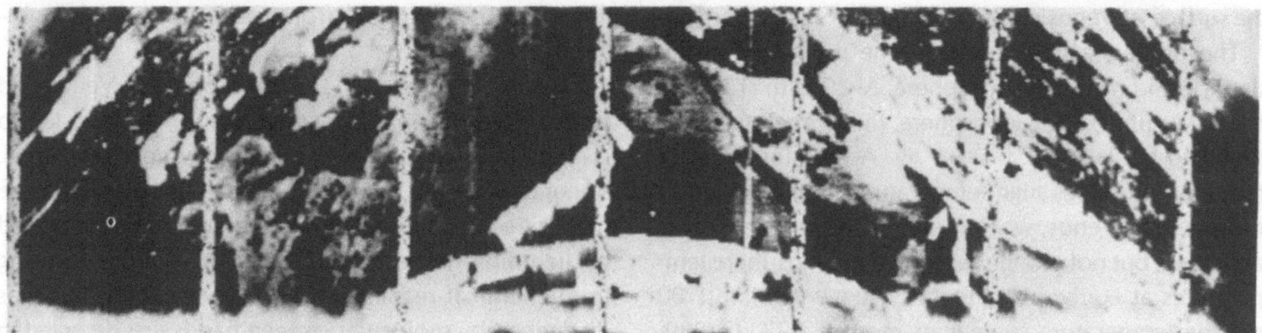

Figure 3.7. Panorama from Venera 10 shows a different type of landscape in which the rocks are more weathered and there is more debris between them.

At both sites the rocks appear to have been broken from a layered mass. The steep slope at the Venera 9 site suggests that there might have been active geological processes at work on Venus, possibly mountain building. Radio telescopes had observed features on Venus rising several kilometers above surrounding terrain. However, observation from orbit had indicated that the surface of Venus is extremely flat—i.e., close to perfectly spherical—which would seem to indicate that there have been no tectonic plate movements of the type associated with mountain building on Earth.

Meanwhile the two orbiters were returning equally important data about the planet. It was when the orbiters moved out of range of the landers that the transmission from the surface was cut off. The landers themselves did not fail. The orbiters transmitted information about the planet and its induced magnetosphere for many months, the first synoptic measurements of the space environment of Venus.

This orbital information established beyond doubt that the clouds of Venus were not composed of water droplets, but that they did, nevertheless, consist of liquid drops. Spectroscopy reinforced the polarimetric studies by showing that there could not be more than 5 parts of water vapor in 10,000 parts of the atmosphere. Moreover, the temperature of the upper clouds was measured as − 40° C (− 40° F) which would preclude the existence of water droplets there. The exact nature of the droplets would not be known until much later and would be discovered from terrestrial-based observations from high flying aircraft of NASA.

Information from the orbiters allowed Soviet scientists to construct a map of the infrared radiation from the cloud layer and showed that the night side is warmer than the day side because radiating material is carried to higher levels by powerful convection streams arising in the daylit hemisphere.

A series of plasma and magnetic field measurements made by the orbital spacecraft of Veneras 9 and 10 allowed a detailed study of the pattern of solar wind flow around Venus. A plasma tail was discovered with some features similar to those of the magnetic tail of the Earth. Bundles of magnetic field lines in the tail were separated by a layer similar to the neutral sheet of Earth's magnetotail. A tail boundary appeared to separate two types of plasma—that arising from the planet and that originating in the solar wind. The discovery of this tail was surprising and led some to estimate that the magnetic field of Venus must be greater than that eventually measured later by the spacecraft. Observations made simultaneously from the two Venera orbiters showed that the measured magnetic field is not an intrinsic field of the planet but arises from induced currents flowing in the convective ionosphere of Venus. The currents are induced by the magnetic field carried by the solar wind streaming past the planet. The magnetic field was found to be even less than that estimated from earlier Venera experiments. The intrinsic value of the field at the surface of Venus could not be greater than 5 gamma.

The orbiters also investigated the properties of the dayside and nightside ionosphere. At 200 km (125 mi), the lower limit of the ionosphere, the temperature was found to be − 73° C (− 163° F). Absorption of solar radiation produces high temperatures in the upper atmosphere of Venus, where much of the solar energy is absorbed, but not as high as terrestrial atmosphere temperatures at equivalent altitudes. Above 160 km (100 mi) the temperature sometimes reached 527° C (980° F) on the dayside at high solar activity periods. Maximum concentrations of about 5×10^5 electrons per cubic cm. were measured at the 140 km (87 mi) level during the day, but concentrations were 50 times less at night. The daytime ionosphere was compressed inward toward the planet by the solar wind.

The important discovery about the ionopause—the boundary between the solar wind and the planet's ionosphere—was that it is compressed by the solar wind. The heights of the ionopause differed considerably between Mariner 5 and Mariner 10, from which it had been concluded that the pressure of the solar wind is an important determinant of the height of the ionopause. The solar wind at the time of Mariner 10's flyby was exerting more pressure than at the time of Mariner 5's encounter with Venus, and the height of the ionopause was 350 km and 500 km (217 and 311 mi) respectively at the flybys.

Data from Veneras 9 and 10 confirmed the dependence on solar wind pressure and also showed that the ionopause depends on solar zenith angle, being lower on that part of the planet directly facing the Sun—only 250 to 280 km (155 to 174 mi). With increasing distance from the subsolar point, the boundary between the ionosphere and the solar wind becomes unstable because of viscous interaction between the two plasmas, instability, and dissipation of energy. A thickening boundary layer develops around the sides of the planet.

The spacecraft confirmed earlier observations by the Mariners that the nighttime ionosphere is very irregular. The Veneras had discovered the magnetic tail of Venus, and identified the induced nature of the magnetic field measured near the planet. They determined the position of the shock front and found that it is asymmetrical.

Existence of clouds in the atmosphere of Venus and study of the dynamic processes occurring there led to speculation that electrical storms may arise on Venus. Veneras 9 and 10 detected short-lived glows on the night side which were later interpreted as being caused by lightning flashes. A constant night airglow arising from chemical reactions in the upper atmosphere was evidenced by molecular oxygen bands first detected in the data from the Venera 9 and 10 orbiters.

There were many questions that still remained unanswered despite the successes of Veneras 9 and 10. The Soviet program continued to develop spacecraft to explore Venus, and although no missions were sent to the planet at the 1976–77 opportunity, two more spacecraft were being made ready for launch at the 1978 window. For this opportunity the U.S. also had two Venus spacecraft ready, Pioneer Venus Orbiter and Pioneer Venus Multiprobe. These American spacecraft are described in the next chapter.

The Soviet spacecraft were delayed by one launch opportunity to give the scientists time to assess the enormous amount of new data returned by Veneras 9 and 10. Moreover, now that the nature of Venus's surface environment was more fully understood the mission planners wanted an improved landing craft. At this time, too, there were discussions with French scientists on the possibility of incorporating a French instrument package on balloons that would be designed to float in the atmosphere of Venus for several days. This Soviet-French mission did not materialize for the 1978 launch window but was deferred until the 1983 window. Later, however, political difficulties caused its further deferment until later in the 1980s.

On September 9, 1978 and September 14, 1978, two more large rockets inserted spacecraft into Earth orbit from the Soviet planetary launch facility at Tyuratam. A short while later Veneras 11 and 12 left the orbital launch pads and started on their interplanetary missions. These huge spacecraft each weighed as much as 5000 kg (11,000 lb), but neither included an orbiter. Each consisted of a flyby bus and a greatly improved lander. A flyby mothercraft was chosen to allow data to be relayed from the surface for longer than was possible via an orbiter. Instrumentation had also been improved and increased in scope. A new gas chromatograph sought more precise measurements of carbon dioxide, carbon monoxide, nitrogen and argon. Since lightning had been suspected, a new instrument was included to try to detect electrical discharges in Venus's atmosphere. Improved cameras were also carried to get more panoramas of the surface.

The flyby spacecraft were equipped as previous spacecraft to gather data about the solar wind and magnetic fields and particles. In addition they acted as extended communications relays for the landers.

Again the landing spacecraft were cooled before entry. They used circular air brakes and parachutes; the air brake was sufficient for the lower regions of the atmosphere—below about 40 km (25 mi). Venera 12 arrived on December 21, 1978; Venera 11 on December 25, 1978. Both entered successfully and landed safely on the surface after taking about one hour to descend through the atmosphere. Venera 11 transmitted data from the surface for 95 minutes, Venera 12 for 110 minutes. In both cases communication ended when the flyby bus flew out of communication range.

Venera 11 landed at latitude 14° S, longitude 299°, and Venera 12 at latitude 7° S, longitude 294°—both in the daylit hemisphere. Unfortunately, no pictures were returned from the surface this time. But there was plenty of other data. Intense thunderstorm activity was recorded, with one thunderclap lasting for 15 minutes. The ratio of argon 36 to argon 40 was measured and found to be 200 to 300 times that in Earth's atmosphere.

The relative abundance of isotopes of the noble gases is of particular importance to our understanding of the evolution of planets and their atmospheres. The isotopes fall into two groups: *radiogenic* isotopes are formed by radioactive decay of elements; *primordial* isotopes were present at the formation of the Solar System. From the relative abundance and the absolute concentration of these isotopes, scientists can gain an understanding of the conditions in the nebula from which the planets formed and conditions arising from subsequent events in the evolution of the planet.

The Soviet Veneras carried a mass spectrograph to seek answers to questions about relative abundance by sorting the ions in small samples of gas. To do this the instrument ionized the gas and collected the ions according to their masses. Another instrument to measure constituents of the atmosphere was a gas chromatograph. This used the different degrees of absorption of various gases by porous substances to separate a mixture of gases into its individual components. An optical spectrometer was also carried.

The mass spectrometer operated from 24 km (15 mi) to landing and obtained many samples. In addition to finding the expected carbon dioxide and nitrogen molecules, the spectrometer also showed the presence of various isotopes of carbon, oxygen, and nitrogen, and of neon, argon, and krypton. The main krypton isotopes were similar in concentrations to those in Earth's atmosphere, but the concentration of argon isotopes was very different. The radiogenic isotope, argon 40, and the primordial isotope, argon 36, were found in equal amounts in the atmosphere of Venus, whereas on Earth the radiogenic argon is 300 times more abundant than the primordial argon. The presence of so much argon 40 in Earth's atmosphere might be explained by assuming that Venus derived most of its atmosphere from the nebula of formation, whereas Earth captured relatively small amounts of gas from the nebula but evolved its atmosphere from its own interior.

The gas chromatograph results indicated that there is very little water vapor in the atmosphere of Venus: at altitudes below 24 km (15 mi) there is less than 0.01 percent water vapor. At such a concentration, if the planet's entire water vapor were condensed onto the surface it would form a layer less than 1 cm deep. This extremely low concentration of water could have arisen in at least three ways: (1) Venus could have originally been formed with less water than Earth or Mars; (2) the planet could have been formed with the same amount of water, but the water was dissociated by solar radiation at early stages of Venus's evolution, so that hydrogen escaped into space, and oxygen was bound up in various chemical reactions with the material of the planet; (3) water is present today, but it is bound closely in hydrated minerals capable of existing under the high surface temperatures of the planet.

The two spacecraft carried spectrophotometers which measured the spectrum of the daylight sky and the angular distribution of brightness. A large amount of solar radiation reaches the surface of Venus, but it is not direct sunlight, it is scattered radiation. You could not see the sun from the surface, or even from the base of the clouds. You would just see a lightening of a region of sky toward the sun.

This solar energy reaching the surface confirms that Venus is heated by a greenhouse effect which traps incoming heat and prevents its reradiation into space.

The Venera 12 lander contained the first instrument to measure the composition of the cloud particles within the cloud layers. These particles, collected on special filters, were analyzed with an X-ray fluorescent spectrometer. Above 61 km (38 mi) the most abundant particle consisted of chlorine. The large particles appear therefore to consist of chlorine compounds. Strangely, sulfur was not measured, although Earth-based observations had concluded that the clouds consisted of sulfuric acid droplets. The mystery deepened.

Following analysis of data from the Venera 11 and 12 missions Soviet scientists summarized the major questions still requiring answers. What is the true explanation for the high content of primordial gases? Why is there so little water vapor in the atmosphere of Venus? What is the true nature of the cloud particles? What is the mechanism responsible for the four-day circulation pattern of the high atmosphere? How active is the interior of the planet? Is there currently volcanic activity on Venus? How long have the present high surface temperatures existed; were they present from immediately after formation, or did they develop more recently?

The Soviets did not send any spacecraft to Venus at the 1980 opportunity; instead they waited for 1982 and in the interim they continued to develop the Venera spacecraft. Venera 13 was launched October 31, 1981 for arrival at Venus on March 1, 1982. Its sister ship, Venera 14, was launched November 4, 1981 and reached Venus on March 5, 1982. The bus spacecraft were designed for flyby missions, the landers for soft landings and transmissions of data from the surface.

After initial slowdown on entry, each lander released a parachute at 65 km (40 mi) altitude. At 150 m/s (492 f/s) a second parachute opened and scientific data were transmitted. At 48 km (29 mi) the parachute was jettisoned and the capsule fell freely, slowed by aerodynamic braking. The capsule landed at about 8 m/s (26 f/s).

Venera 13 landed at 7.5° S latitude, 303° longitude, Venera 14 at 13.4° S latitude, 310° longitude—both in the daylit hemisphere but at geologically different sites. Venera 13 was in rolling foothills east of a small highland area named Phoebe. Venera 14 was on flat plains. Both landings were successful and good data were returned. Instruments were switched on about the time the descent parachutes opened, and they sent data for an hour until the landing. On the surface data transmission continued for a further 2 hours from Venera 13, and 57 minutes from Venera 14.

A mass spectrometer of greater sensitivity was carried on each of these planetary probes and it provided data on the minor constituents of the atmosphere. The ratio of the neon isotopes, neon 20 to neon 22, was measured as 11.8, somewhat less than what had been determined earlier from Pioneer Venus measurements. It was still apparent that the ratio on Venus was a little higher than that of the terrestrial atmosphere but lower than that of the solar wind. However, results for krypton on Venus were still confusing. Considerable differences were found from earlier measurements.

Another puzzle was the detection by the Veneras of carbonyl sulfide (COS) and hydrogen sulfide (H_2S) neither of which were detected by an instrument carried by Pioneer Venus probes. For some time theoreticians had assumed that COS was an important molecule in the chemistry of the clouds. That it was not found by Pioneer Venus had led to reevaluation of several theories. The new measurements by the Veneras 13 and 14 compounded the mystery.

Water vapor also varied considerably from other attempts to measure its ratio. The data from the Veneras suggested that water vapor was being removed from the atmosphere within the clouds and released at the cloud base. The absorption in the clouds could be explained by the known presence of droplets of sulfuric acid. However, another mystery was why the amount of water vapor decreased in the lower atmosphere progressively toward the surface. Some other absorber was postulated, but its nature was unknown.

Wind velocity at the surface was measured by acoustic sensors instead of by the anemometers carried on earlier Veneras. The surface wind was estimated as blowing at less than a half meter per second.

New information was obtained about the clouds, and the presence of fine structure at the bottom of the lower cloud layer, as suspected from Pioneer Venus data, was confirmed. Aerosol particles were collected during the descent of the Venera probes and the trapped particles were examined by X-ray fluoroscopy. Sulfur was the most abundant element, as expected theoretically. Chlorine was also present, but in smaller amounts than the sulfur, again as expected. However, the opposite result had been obtained from the Venera 12 measurements. Why two similar instruments should give such differing results remained a mystery.

The success of earlier spacecraft in obtaining the first images of the Venus surface was repeated with Veneras 13 and 14, and actually improved upon. This time the cameras provided a full 360° field of view and were capable of producing some images in color. Venera 13, in its two hours of operation on the surface, obtained eight images. The colored pictures show a brownish surface. A smaller pixel size also improved the resolution of detail. Again the most prominent objects in the pictures were the large flat rocks, many with sharp edges and a layered structure, with dark soil between them—different from the rocks shown in the Venera 9 and 10 pictures. The pictures look very similar to pictures of the ocean floors of Earth, perhaps because at the high pressures involved on Venus sedimentation of debris might be analogous to the way sediments collect on the ocean floors of our planet.

The composition of the rocks was analyzed by way of soil samples. The Veneras carried a device which about half a minute after each landing drilled into the soil and in about 12 minutes collected small (1 cc) samples that were then taken into the lander for analysis. The sample passed through a series of chambers of decreasing pressure. The X-ray fluorescence spectrometer showed that the soil consists of fragments of quite dissimilar rocks at the two sites. Venera 13 was in an area where there was a high percentage of potassium in a basaltic rock that is rare on Earth. Venera 14 was in an area of rocks similar to terrestrially common tholeiitic basalt.

As a result of the aggressive Soviet exploration of Venus, data were being gradually amassed on the different regions of the planet which was being shown to be a complex world that might have been Earth-like earlier in its history. These interpretations of the large amount of new data from the planet are discussed more fully in the final chapter of this book.

Meanwhile, two more Soviet spacecraft were being readied while U.S. scientists were trying to convince their government to fund a Venus radar mapper for the mid-1980s with varying degrees of success. Approval was finally obtained for the mapper to be sent to Venus in 1988, but there were no American plans for landing spacecraft on Venus.

On June 2, 1983, Venera 15 left Earth orbit for Venus, followed, on June 7, by Venera 16. These were huge automated spacecraft designed as radar mappers, ahead of the proposed U.S. mission by some 5 years and improving considerably in resolution if not in areal coverage on the radar mapping by Pioneer Venus. Pioneer Venus had a surface resolution, as described in the next chapter, of about 20 km (12 mi). The new Russian Venus radars resolved the surface features to between 1 and 2 km (0.6 and 1.2 mi), a great improvement. This was not as good as the resolution proposed for the U.S. mapping radar spacecraft, but the Venera was already exploring Venus and was not merely a plan for the far future. The two spacecraft arrived at Venus in mid-October 1983 and were successfully placed in identical polar orbits with a period of 24 hours and a periapsis of 1000 km (620 mi) at 60° N latitude.

Within a few days their synthetic aperture radar systems returned high resolution pictures to Earth. Coverage was limited, however, to the northern hemisphere of Venus in 9000 km (5600 mi) scans, each about 150 km (95 mi) wide. These tracks moved eastward at a rate of about 1.5 deg. of longitude per day. The radar images showed impact craters, hills, major fractures, benches, mountain ranges. A variety of geologic phenomena was recognized which indicated to Soviet scientists that there has been a long period of active tectonics operating on the planet. Initially the mapping concentrated on north polar regions, and the Soviets reported discovery of a continent-sized mass of elevated terrain there, about twice the size of Australia with a periphery of very high escarpments.

In the north polar area is a clearly defined volcanic dome surrounded by extensive lava flows. Another large caldera, similar to the calderas of the martian volcanoes on Tharsis, is also recognizable in the polar region. The caldera is about 60 km (40 mi) across. About this same time ground-based radar observations of Venus with the 300-meter (1000-ft) diameter Arecibo radio telescope in Puerto Rico were also revealing craters and folded terrain resembling the Appalachian Mountains. The radio telescope images were of comparable resolution to the Venera images, but they covered a smaller region of the planet than that which can be observed from orbit. These radar images confirmed what had been earlier suspected, that the Beta Region of Venus is one of large volcanoes and lava flows. Also, in that area, there is a large canyon comparable in size with Arizona's Grand Canyon. Another image showed a large crater in Maxwell, which is a mountainous region of Ishtar Terra, a "continent" of Venus.

The next stage in the Russian conquest of Venus is the use of a more advanced interplanetary spacecraft named Vega. First launches of Vega took place successfully in December, 1984 with a backup spacecraft held in reserve should one or both fail. The two complex spacecraft will reach Venus in June 1985. As each bus flies past Venus it will release a lander probe and a balloon. The lander will descend to the surface as with previous missions, the balloon will float in the atmosphere and be tracked for over two days to provide data on motions of the atmosphere. Each balloon will also measure vertical winds and temperature and pressure. Each balloon has a diameter of about 5.4 meters (18 ft).

The landers will continue with the same types of measurements as earlier landers of the Venera series. They will also analyze samples of surface materials. The flyby bus will zoom past Venus and continue to an encounter with Halley's Comet in March 1986. The comet will be inspected by wide-angle and narow-angle television cameras to reveal detail as small as 200 meters (600 ft) across.

4

PIONEER VENUS

As mentioned in an earlier chapter, the study of concepts for the Pioneer Venus project began in 1967, shortly after NASA's Mariner 5 spacecraft flew by Venus and the USSR's Venera 4 spacecraft probed into the atmosphere of the planet. Three scientists, R. M. Goody at Harvard University, D. M. Hunten at Kitt Peak National Observatory, and N. W. Spencer at NASA's Goddard Space Flight Center, formed a group to look into the feasibility of a simple entry probe to investigate the atmosphere of Venus.

Up to this time ground-based observations had contributed very little to detailed knowledge about the planet, and the few spacecraft that had passed close to Venus had added only a comparatively small amount of new information. Speculation about Venus and its environment was still greater than information about the planet. It was apparent that a different technique of exploration was needed for Venus compared with, say, the Moon and Mars where landings were feasible and the surfaces were visible from Earth and from orbit. The early science study groups who were trying to encourage a national commitment to explore Venus first defined the key scientific questions about Venus, and then attempted to define the types of spacecraft missions needed to seek answers to those questions. By contrast, earlier space missions had been designed mainly on the basis of accepting available space technology, and then selecting scientific experiments that could be attempted by applying that technology in the form of available spacecraft and boosters. To explore Venus effectively with spacecraft it became increasingly apparent that scientists needed to select scientific payloads to attack broad objectives, and engineers needed to develop the necessary new spacecraft to carry those scientific payloads to Venus.

Several approaches to a mission to Venus were considered by a group based at NASA Goddard Space Flight Center. These approaches included a balloon floating in the Venus atmosphere, atmosphere probes, and orbiters (figure 4.1). Early studies also looked into the relative merits of a flyby mission with probe release from the flyby spacecraft, a direct impact bus with separate probes released before the bus reached Venus, and an orbiter from which probes would be released (figure 4.2).

By the end of 1969 NASA had merged various con-

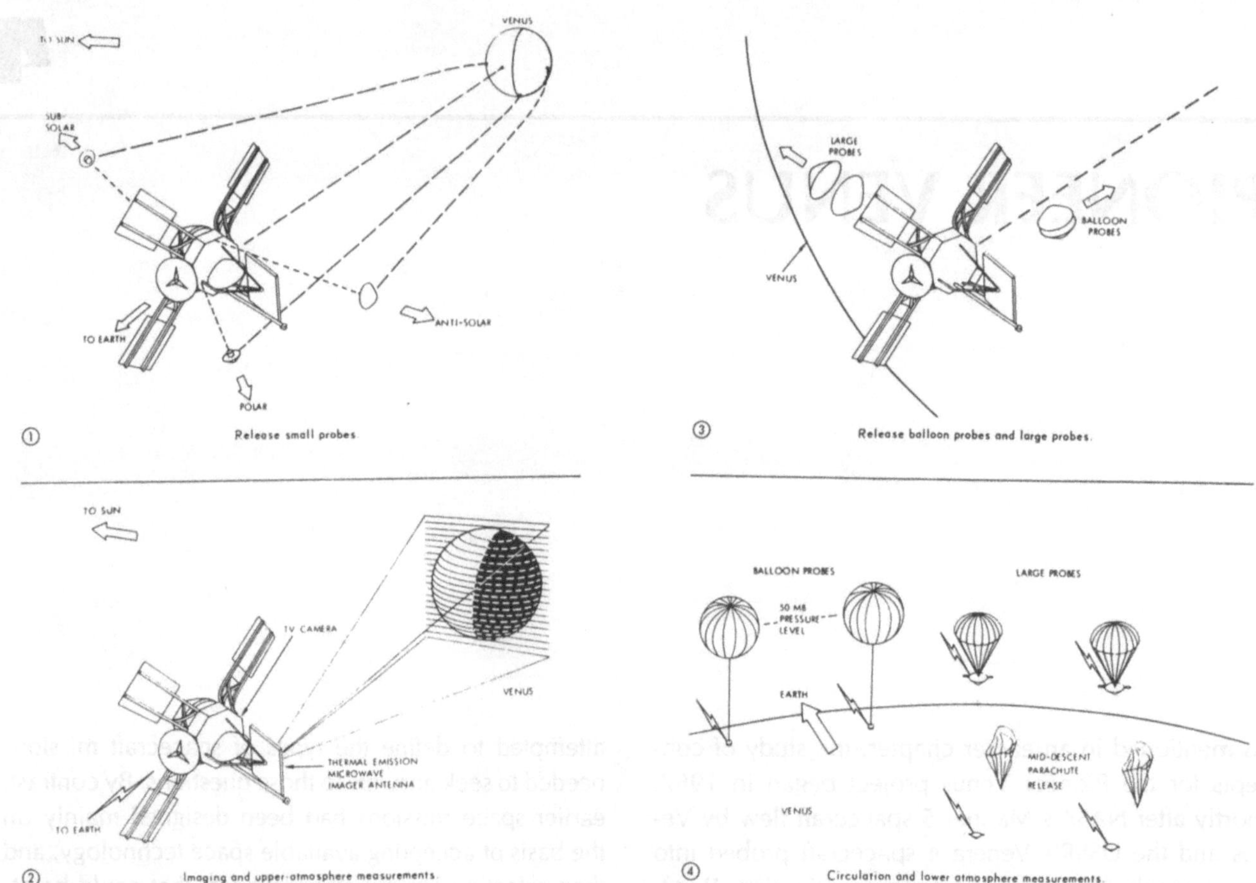

Figure 4.1. There are several ways in which a mission to Venus can be organized, as shown in these diagrams which illustrate an early plan to release probes and balloons from a flyby spacecraft. (NASA/GSFC)

cepts into what became known as the *Universal Bus;* a combination of a Venus probe spacecraft and a Planetary Explorer orbiter spacecraft, both of which had been studied by various groups interested in exploring Venus. The Universal Bus could be used to deliver multiple entry probes into the atmosphere of Venus or to send a spacecraft into orbit around the planet.

In 1970, 21 scientists of the Space Science Board and the Lunar and Planetary Missions Board of NASA looked at the scientific potential of missions to Venus based on the technology amassed from experience with Explorer spacecraft. They issued a report entitled *Venus—Strategy for Exploration, 1970* which was colloquially referred to as the "Purple Book" after the color of its cover. As mentioned in an earlier chapter this report recommended that exploration of Venus should be prominent in the NASA program for the 1970s and 1980s, and that the Delta-launched, spin-stabilized Planetary Explorer spacecraft should be the main vehicle for initial missions using orbiters, atmospheric probes, and landers. Also the report said that to keep the cost of these missions down, NASA should be prepared to take some higher risks than the agency had accepted for earlier space missions.

The strategy to explore Venus assumed optimistically that one or two launches would be attempted at each launch opportunity. In contrast with Soviet strategy, hybrid missions (e.g. a spacecraft carrying both an orbiter and an atmospheric probe) were not recommended. Two multiprobe missions were proposed for

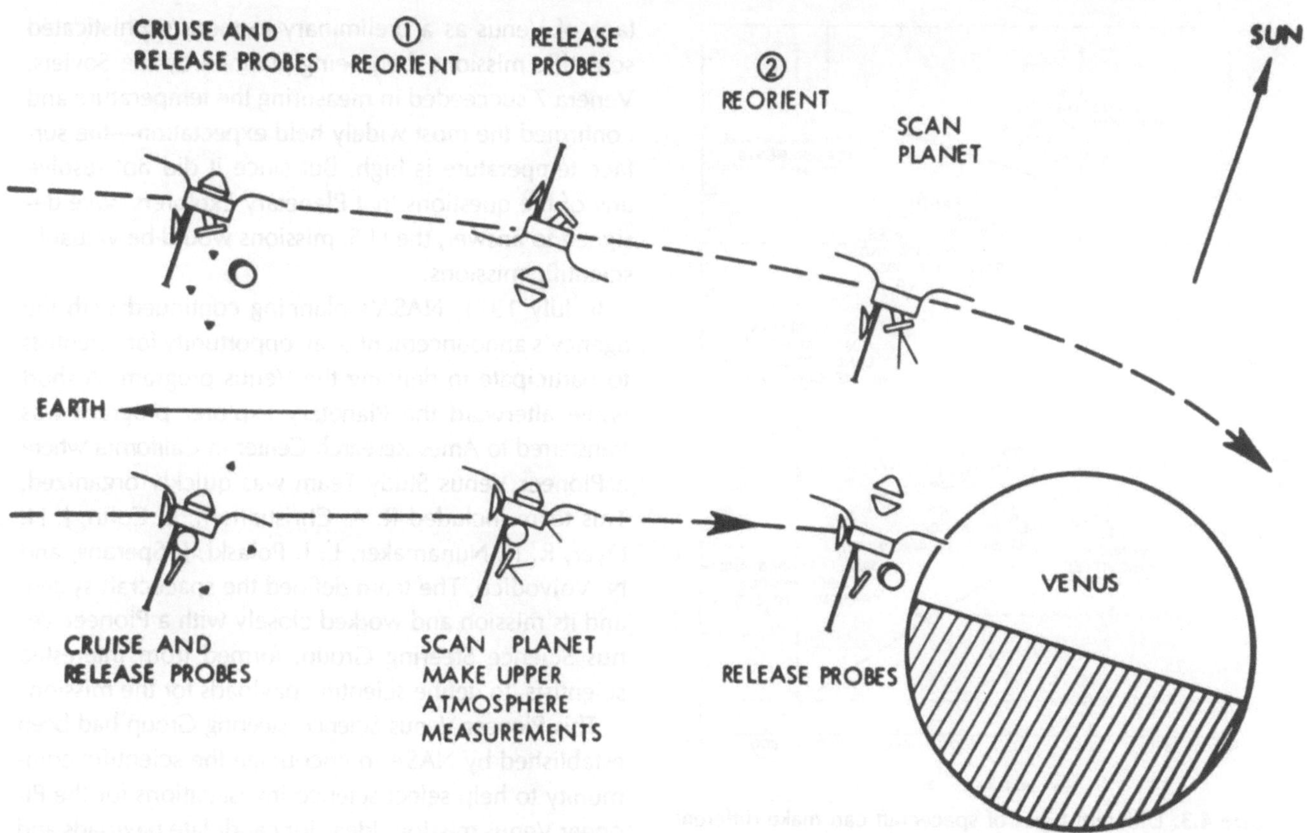

Figure 4.2. Probes can be released from a flyby spacecraft or from a bus which itself also penetrates the atmosphere. The Soviets used hybrid missions, but the U.S. opted for separate specialized spacecraft as being more reliable. (NASA/GSFC)

the 1975 opportunity, and two orbiters for the 1976–77 opportunity. The use of orbiters, landers, and balloons at subsequent opportunities was not specifically defined, but a landing mission was suggested for the 1978 opportunity.

An in-depth study of Venus, continued the report, promised to reveal new insights into the evolution of planets. Because of the opaque atmosphere of Venus, which screened its surface, and the absence of a satellite, which prevented accurate calculation of mass and shape, knowledge of Venus in 1970 was not comparable to that of Mars. Scientists needed other important information about Venus such as the chemical composition and minerology of the surface materials, heat flux from the interior, presence or absence of an iron-rich core, and the seismic qualities of the planet. It was recognized that the high surface temperature would make it extremly difficult to place instruments on the planet's surface to obtain this information. Nevertheless, U.S. scientists agreed that a program of measurements on a scale proposed for the Planetary Explorers could make highly significant measurements from orbiters, probes, and landers (figure 4.3). For example, surface elevations could be measured from orbit with a radar altimeter.

However, in the fall of 1970 politics were such that it was impossible to fund a new program for planetary exploration that could meet a launch date suitable for the 1975 opportunity. New starts for planetary programs were extremely difficult to move through Congress. As a consequence the entire American program for the exploration of Venus slipped badly. The NASA

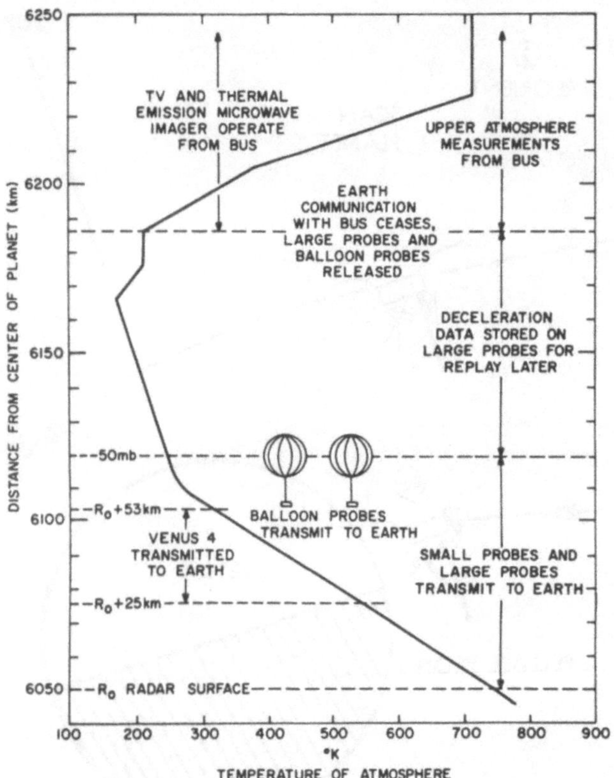

Figure 4.3. Different types of spacecraft can make different types of measurements in Venus's environment, as shown on this diagram. The regions for study by a bus, a probe, and a balloon are identified. (NASA/GSFC)

plan had to be revised to schedule two multiprobe spacecraft for launching during the 1976–77 opportunity, followed by a single orbiter spacecraft in 1978, and a single multiprobe (consisting of a floating balloon probe and a lander) in 1980.

Soon after this report appeared, the Soviet Venera 7 spacecraft, described in the previous chapter, successfully entered the atmosphere of Venus. Questions arose as to whether the recommendations of the 1970 study still applied. A special panel of experts reassessed the study and confirmed the earlier recommendations. They said that the American Planetary Explorer program should still have a high scientific priority. It was pointed out that Venera 7 was a highly specialized probe designed to perform onlv two functions—to measure atmospheric temperature and pressure down to the surface of Venus as a preliminary to more sophisticated scientific missions then being planned by the Soviets. Venera 7 succeeded in measuring the temperature and confirmed the most widely held expectation—the surface temperature is high. But since it did not resolve any of the questions that Planetary Explorers were designed to answer, the U.S. missions would be valuable scientific missions.

In July 1971, NASA's planning continued with the agency's announcement of an opportunity for scientists to participate in defining the Venus program. A short while afterward the Planetary Explorer program was transferred to Ames Research Center in California where a Pioneer Venus Study Team was quickly organized. This team included R. A. Christiansen, L. Colin, J. H. Dyer, R. R. Nunamaker, L. J. Polaski, J. Sperans, and N. Vojvodich. The team defined the spacecraft system and its mission and worked closely with a Pioneer Venus Science Steering Group, formed from interested scientists, to define scientific payloads for the mission.

This Pioneer Venus Science Steering Group had been established by NASA to encourage the scientific community to help select science investigations for the Pioneer Venus mission. Ideas for candidate payloads and spacecraft were developed to be used as guides by a Payload Selection Committee and by contractors who would later have to design and develop them. The Science Steering Group held several meetings, and in June 1972 it published a report which became the accepted U.S. guide to Venus exploration. Colloquially referred to as the "orange book"—again because of the color of its cover—the *Pioneer Venus, 1972* report reviewed and endorsed the scientific rationale for missions to Venus. It took into account developments since the earlier U.S. reports, including the delay in starting the program and the scientific results obtained from Venera 7, from new Earth-based observations, from new theoretical analyses, and from continuing analysis of data gathered by earlier Soviet and American spacecraft. The report recommended that the missions to Venus should be supported and should consist of a multiple probe in 1976–77, a single orbiter in 1978, and a probe mission in 1980.

The Science Steering Group's report emphasized that

most of the scientific questions concerning Venus required measurements within the atmosphere and below the cloud tops to as close as possible to the surface. It was still doubtful at this time that spacecraft could be designed and fabricated with the ability to survive on and return data from the hot surface of the planet. A probe mission, at least into the atmosphere of Venus, if not to the surface, was thought to be desirable at the first opportunity. Two launches of identical spacecraft were recommended in case of a failure. If both launches were successful and the first probe gathered useful data, a second spacecraft arriving later at Venus could be retargeted at the planet on the basis of experience with the first.

The study described the mission as consisting of two identical spacecraft and payloads despatched to Venus during the launch opportunity from December 1976 through January 1977. Each spacecraft would consist of a bus, carrying one large and three small probes. The spacecraft would be spin stabilized and would derive electrical energy from solar radiation. Interplanetary cruise from Earth to Venus would last about 125 days, and the probes would be separated from the bus about 10 to 20 days before entry. The two large probes would each be equipped with an aeroshell to smooth entry into the atmosphere, a heat shield to protect against the heat of entry, a parachute system, and a pressure vessel to protect the experiments during the descent into the dense, hot, lower atmosphere. Six identical small probes would each consist of an aeroshell, a heat shield, and a pressure vessel. But these probes would fall freely to the surface without using a parachute. The two buses would also enter the atmosphere at a shallow angle and transmit data about the upper atmosphere until they burned up.

For the next opportunity in 1978 the mission would be to place a spacecraft in orbit around Venus. The orbiter would be spin-stabilized and it, too, would obtain electrical power from solar cells. It would be launched during the period from May to August 1978. After its interplanetary cruise, the length of which would depend upon the actual launch date, the spacecraft would be placed in an elliptical orbit about Venus to gather data for at least one Venus day. The study recommended another probe in 1980, but did not make any recommendations for a mission in 1982.

It is important to note that at this time, despite Russian entry probes and flybys, we still knew very little definitive about the lower atmosphere of Venus. For example, we did not know the number and thickness of the cloud layers, nor their composition. At least three different hypotheses were current to account for the high surface temperature of the planet. In support of the missions to Venus scientists cited the important need for data to unravel the puzzle of the origin and evolution of the Solar System and planets in general, and the formation, evolution, and current dynamics of planetary atmospheres.

After an independent study of the Soviet Venus program, the Science Steering Group agreed with the Space Science Board's earlier assessment of the Venera Program. The previous eleven years of Soviet exploration of Venus had produced measurements within the lower atmosphere, such as pressure, temperature, density, and gross atmospheric composition. However, the National Academy of Science's Venus Study pointed out that there was still a wide range of scientific problems concerning the magnetosphere, the upper atmosphere, the lower atmosphere, and the solid planet that had not been addressed by the Soviet program and were most important to deriving an acceptable explanation of conditions on Venus compared with those on Earth and Mars.

In recommending the types of instruments to be carried by the spacecraft, the Science Steering Group adopted a conservative approach to avoid high costs because the nation was still balking at making any significant effort to explore other planets. The Steering Group suggested that an instrument should be known to have performed successfully in Earth's atmosphere before its selection for the mission to Venus, and that no new or untried concepts of measurements should be attempted. Wherever possible, instruments should have been qualified for use in spacecraft or aircraft, and if not, it was essential that the instrument be so simple and rugged that its satisfactory performance could be readily assured by laboratory tests beforehand.

Charles F. Hall had led a team at NASA Ames Research Center for many years after he won approval from NASA's deputy director for space sciences to develop an Interplanetary Pioneer spacecraft in 1962. He managed the Pioneer program starting with Pioneer 6, which was launched into solar orbit in 1965, and continued through the highly successful Pioneers 10 and 11, which were the first spacecraft to reach Jupiter and Saturn and later to escape from the Solar System. Pioneers 6 through 11 were built by TRW, Inc. The two Pioneer Venus spacecraft were the culmination of this successful line. They were built by Hughes Aircraft Company with Ames Research Center continuing in the role of project management and with Charles Hall as the project manager. Later Richard O. Fimmel was project manager for the extended missions of all the Pioneers.

The Pioneer Venus mission finally emerged from the planning studies as a Multiprobe and an Orbiter, both of which incorporated significant and major advances in sophistication of spacecraft and their instrumentation compared with earlier spacecraft used to explore Venus. The new space program was initiated as an example of a model low-cost program based on innovative approaches to management and an understanding that the total cost would be kept below $250 million, which represented about 1 dollar per head of the U.S. population.

The aim of the Pioneer Multiprobe was to obtain significant information on, or make important contributions to answering questions about: cloud layers, forms, physics, and composition; solar heating of the atmosphere, its deep circulation and driving forces; loss of water, stability of carbon dioxide, and vertical temperature structure; ionospheric turbulence, ion chemistry, exosphere temperature, magnetic moment, bulk atmospheric composition; and the amenopause where the solar wind reacts with the planet's atmosphere.

The Pioneer Orbiter would do likewise in respect to cloud forms, extent of the four-day circulation, cause of the four-day circulation, loss of water, vertical temperature structure, ionospheric motions, ion chemistry, exospheric temperature, bulk atmospheric composition, the amenopause, topography, magnetic moment, and gravitational moments.

To formulate a plan for obtaining such information, the Science Steering Group defined important questions about Venus that could be investigated. First, about the atmosphere; the scientists wanted answers to questions concerning the atmosphere in general. Where is solar radiation deposited within the atmosphere? What are the major gases in the atmosphere? How do they vary at different altitudes? Why is carbon dioxide stable in the upper atmosphere? How does the four-day circulation vary with latitude on Venus and depth in the atmosphere and does it vary over the planet? How much turbulence is there in the deep atmosphere of Venus? What is the nature of the wind in the lower regions of the atmosphere? Is there any measurable wind close to the surface? What are the horizontal differences in temperature in the deep atmosphere that act as driving forces there? What are the horizontal temperature differences at the top layer of clouds that could cause the high winds there and account for the four-day circulation? Is there an isothermal region in the atmosphere or other departures from adiabaticity? What is the structure of the atmosphere near the tops of the clouds?

There were also questions about the ubiquitous clouds. How many cloud layers are there in the atmosphere of Venus, and where are they located? Do they vary over the planet? What are the forms of cloud, and are the clouds layered, turbulent, or merely hazes? Are the clouds opaque? What are the sizes of the cloud particles? How many particles are there per cubic centimeter? What is the chemical composition of the clouds? Is it different in the different layers?

Another important area of investigation concerned the upper atmosphere and how the solar wind interacts with the planet. Does the planet have any internal magnetism to hold off the solar wind? What is the temperature in the exosphere and does it vary over the planet? What is the chemistry of the ionosphere? Are ionospheric motions sufficient to transport ionization from the day to the night hemisphere?

The nature of the surface of Venus posed many more questions of importance to comparative planetology. What features exist on the surface of the planet? How do they relate to thermal maps? What is the composition of the crustal rocks of Venus? Is there any seismic activity and if so what is its level? Has water been lost from Venus? If so, how? Do tidal effects from Earth exist at Venus, and if so, how strong are they? What is the figure of the planet? What are the higher gravitational moments? Are plate tectonics active on Venus? How has the surface evolved differently from other terrestrial planets?

Early in 1972 the European Space Research Organization (ESRO) stated it was interested in becoming involved in the orbiter mission planned for 1978. NASA planned to produce the Orbiter version of the basic spacecraft and provide this to ESRO together with some equipment common to the Orbiter and the Multiprobe. ESRO would then adapt this bus as needed for the orbital mission, and would provide the retromotor to inject the spacecraft into orbit around Venus. Also ESRO would provide a high-gain antenna for communications at high data rates, and would be responsible for integration of scientific experiments. Finally, ESRO would undertake qualification tests on the spacecraft and its payload before delivering the Orbiter to NASA for launch and flight operations.

After a period of in-depth study, a Joint Working Group of European and U.S. scientists issued a report entitled *Pioneer Venus Orbiter, 1973*. This report stated that the Orbiter should operate in orbit for at least one sidereal year of 225 Earth days and preferably for one period of rotation of Venus, namely 243 Earth days. The report defined experiments and the characteristics of the science instruments to measure four important areas of interest regarding Venus. Interaction of the solar wind with the ionosphere would be investigated by a magnetometer, a solar wind and photoelectron analyzer, and an electron and ion temperature probe. Aeronomy, which covers atmospheric composition, photochemistry, and the airglow, would be investigated by a neutral mass spectrometer, an ion mass spectrometer, and an ultraviolet spectrometer/photometer. The atmosphere's thermal structure and density would be investigated by an infrared radiometer and a dual frequency (S- and X-band) occultation experiment. Finally, the surface topography, reflectivity, and roughness would be investigated with a radar altimeter.

Scientists were extremely interested in determining the gravitational field and geometrical shape of Venus. Such information is important to understand the origin and evolution of the inner planets of the Solar System. Gravitational experiments require that an orbital spacecraft should have a periapsis high enough to avoid any atmospheric drag and also be capable of remaining in orbit for sufficient time to amass many data points of tracking. Since there was a conflict between *in situ* measurements, requiring a low periapsis, and gravitational measurements, requiring a high periapsis, the Working Group recommended that the mission should be extended beyond the nominal 243 days. In a nominal mission the periapsis would be low enough for measurements within the atmosphere, then the periapsis would be raised to make accurate gravity measurements. Unfortunately, despite their major contributions to defining a mission to Venus, the Europeans decided later not to participate in the Pioneer Venus program, but many of their recommendations were followed.

Because of tight budgets NASA decided in August 1972 to restrict the program to two flights only, a Multiprobe at the first opportunity in 1977, and an Orbiter at the second opportunity in 1978.

An announcement of opportunity for scientists to participate in the mission of the Multiprobe was issued by NASA in September 1972. As well as investigators who would be responsible for developing the hardware for the science instruments, interdisciplinary scientists and theoreticians were invited to participate. By August 1973 a preliminary payload for the Multiprobe mission had been selected.

The announcement of opportunity for the Orbiter mission was not issued until August 1973. During the ensuing months a NASA Instrument Review Committee evaluated the instrument design studies for the Multiprobe mission and the proposals for scientific payloads for the Orbiter mission. Recommendations were made to NASA Headquarters in May 1974, and the payloads were finally selected on June 4, 1974.

Twelve instruments were chosen for the Orbiter, seven for the Large Probe, three identical instruments for each of three Small Probes, and two for the Multiprobe Bus. In addition several radioscience experiments were chosen that were applicable to all spacecraft (table 4.1).

The big challenge in developing a space mission is always that of meeting the inexorable schedule of the launch date. There were times when Pioneer Venus ap-

Table 4.1. Science Payloads, Their Acronyms, and Principal Investigators[a]

Composition and Structure of the Atmosphere
Large Probe Mass Spectrometer (LNMS), J. Hoffman
Large Probe Gas Chromatograph (LGC), V. Oyama
Bus Neutral Mass Spectrometer (BNMS), U. Von Zahn
Orbiter Neutral Mass Spectrometer (ONMS), H. Niemann
Orbiter Ultraviolet Spectrometer (OUVS), A. Stewart
Large/Small Probe Atmosphere Structure (LAS/SAS), A. Seiff
Atmospheric Propagation Experiments (OGPE), T. Croft
Orbiter Atmospheric Drag Experiment (OAD), G. Keating
Clouds
Large/Small Probe Nephelometer (LN/SN), B. Ragent
Large Probe Cloud Particle Size Spectrometer (LCPS), R. Knollenberg
Orbiter Cloud Photopolarimeter (OCPP), J. Hansen
Thermal Balance
Large Probe Solar Flux Radiometer (LSFR), M. Tomasko
Large Probe Infrared Radiometer (LIR), R. Boese
Small Probe Net Flux Radiometer (SNFR), V. Suomi
Orbiter Infrared Radiometer (OIR), F. Taylor
Dynamics
Differential Long Baseline Interferometry (DLBI), C. Councelman
Doppler Tracking of Probes (MWIN), A. Kliore
Atmospheric Turbulence Experiments (MTUR/OTUR), R. Woo
Solar Wind and Ionosphere
Bus Ion Mass Spectrometer (BIMS), H Taylor
Orbiter Ion Mass Spectrometer (OIMS), H. Taylor
Orbiter Electron Temperature Probe (OETP), L. Brace
Orbiter Retarding Potential Analyzer (ORPA), W. Knudsen
Orbiter Magnetometer (OMAG), C. Russell
Orbiter Plasma Analyzer (OPA), J. Woolfe
Orbiter Electric Field Detector (OEFD), F. Scarf
Orbiter Dual-Frequency Occultation Experiments (ORO), A. Kliore
Surface and Interior
Orbiter Radar Mapper (ORAD), G. Pettengill
Orbiter Internal Density Distribution Experiments (OIDD), R. Phillips
Orbiter Celestial Mechanics Experiments (OCM), I. Shapiro
High Energy Astronomy
Orbiter Gamma Burst Detector (OGBD), W. Evans

[a]Each experiment involved teams of scientists under the leadership of those named above

peared to be progressing two slowly, and there were concerns about meeting the schedule. But all the science instruments for the mission were ready on time. There was one concern about the infrared radiometer when it ran into significant development problems in the detector array. However, the instrument was built and tested in time for the launch.

Although several instruments overran estimates of cost the overall science program cost much less than budgeted, because other instruments cost less than expected. Many of the experiments were indeed unique. There was no suitable off-the-shelf hardware that could be used as originally recommended, and a major challenge was to package all the instruments into a small pressure shell that could withstand the journey through the hostile environment of Venus.

One example of innovative approach was that used to solve the problem of obtaining a suitable neutral mass spectrometer for operation in the hot, dense atmosphere of Venus. A major task connected with the mass spectrometer was to develop an inlet system that would operate over a wide range of atmospheric pressures (up to 100 times Earth's sea level pressure) and still maintain the very low pressure required within the instrument for it to operate. The inlet had to admit very small quantities of gas, yet enough for analysis before being purged ready to accept a subsequent sample. The design selected relied upon a ceramic microleak inlet which took two years to develop. Although this inlet was extremely difficult to develop it worked very well during the mission.

Another difficulty was that although a single inlet could admit a large enough sample of gas in the dense lower atmosphere, an additional inlet was required to provide a sufficient sample in the rarefied upper atmosphere. The second inlet was left open until about the time of parachute release, when a pyrotechnic device crushed the line and stopped further entry of gas. Even if the cutoff device failed there would still be useful data although somewhat degraded.

Scientists were also concerned about what would happen when the spacecraft went through the clouds of Venus where sulfuric acid droplets might condense on the inlet and block it. A heater coil was installed around the inlet to counter this by vaporizing any such condensation.

This instrument also used the first microprocessor to be flown in space, an Intel 4004. The microprocessor permitted a full range of atmospheric composition data to be sampled once every minute over the range covered by the instrument. Without the microprocessor, a full sampling of the atmosphere could be transmitted only once every 10 km (6 mi) of altitude change. With it, the sampling rate was ten times better.

Other instruments also had problems. NASA Ames Research Center had developed a gas chromatograph for the Viking landings on Mars. Experience gained with this instrument was applicable to an instrument for Pioneer Venus. Originally the instrument was considered a backup instrument; one that would provide some details of atmospheric composition should the mass spectrometer fail, but in fact the two instruments complemented each other. The gas chromatograph could measure water vapor which could not be measured reliably by the mass spectrometer and such a measurement was very important. The gas chromatograph was also essential to helping to resolve a scientific controversy that had been developing about the amount of various isotopes of argon in the atmosphere of Venus which has important implications as to how Venus's atmosphere originated and evolved.

A cloud-particle-size spectrometer also proved to be an interesting instrument to develop for the probe. Robert Knollenberg, a cloud physicist, had developed a small instrument that was being used by the U.S. Air Force to measure the number of ice particles in clouds. This seemed a great way to get information about the particles in the clouds of Venus, so the instrument was further developed for the Pioneer Venus mission by Knollenberg and Ball Brothers. Essentially it was an optical bench with a laser at one end and a prism at the other. The difficulty was that part of this bench had to be outside the pressure hull of the spacecraft. Design problems were encountered in safeguarding it against twisting and other distortions that would be expected as the pressure vessel heated in the atmosphere of Venus.

Heaters were also required on the window in the pressure vessel and on the prism outside the window. Fabricating the instrument so that a precise optical bench could penetrate a wall that was changing relative to the rest of the optical bench was an enormous technical challenge. Nevertheless, all the problems were solved in time to meet the launch date, and the instrument worked well as a result of much detailed and innovative engineering during its development.

Another unique instrument carried by the probe spacecraft was designed to measure solar flux. This was important to finding out how the clouds selectively absorb solar radiation. The net flux radiometer flown on each of the small probes had a plate that flipped back and forth to look at the upward and downward radiation flux. It also required two diamond windows on each side, and the instrument hung out over the back of the probe. Its strange appearance resulted in its being referred to as "The Lollipop." The diamond windows were cut from the same stone as a big window for an infrared radiometer carried by the Large Probe to make sure that infrared transmission characteristics would be identical and thereby aid correlation of data from the two instruments. Diamond was the only substance available that would transmit infrared radiation and be capable of withstanding conditions in Venus's atmosphere. There were thus two diamond windows for each Small Probe which, together with the single large window in the Large Probe, brought a total of seven diamonds carried to Venus.

For the Orbiter the most significant instrument development was that of the radar mapper. It weighed only about 11 kg (24 lb), consumed a mere 30 watts, and incorporated over 1000 microcircuits. This was the first time a complex instrument for radar mapping had been assembled in such a compact package. The imaging system aboard the Orbiter was a second generation of the imaging photopolarimeter flown on the Pioneer spacecraft to Jupiter and Saturn. It was given an improved telescope and a new interface.

Meanwhile, while scientists developed their instruments to peer into and beneath the veils of Venus, contractors developed the new spacecraft. Three study contracts of $500,000 each were awarded in August 1972 to define the spacecraft system by June 30, 1973. (The awards went to Hughes Aircraft Company's Space and Communications Group, teamed with General Electric Company; to TRW Systems Group, teamed with Martin Marietta; and to AVCO Corporation Systems Division. AVCO decided later to withdraw from the program.)

Two different approaches were developed. TRW suggested the use of different types of spacecraft for the

Bus and the Orbiter. Hughes preferred one dual-purpose spacecraft. Probes were similar in design concept but the Orbiter configurations differed. In the TRW study the spin axis of the Orbiter was aligned parallel to the plane of Earth's orbit and pointed toward Earth. A fixed high-gain antenna was also pointed to Earth, like the design of TRW's Pioneer Jupiter/Saturn spacecraft. In this adaptation for Venus it was intended that several instruments would be mounted on a movable platform so that they could scan the surface of Venus from orbit. The Hughes design took an entirely different approach and placed the spacecraft's spin axis perpendicular to the Earth's orbital plane so that the spinning of the spacecraft swept the field of view of instruments across Venus as TRW's Pioneer spacecraft did for the Jupiter/Saturn mission. The high-gain antenna pointed independently toward Earth on the Hughes concept. After an in-depth evaluation of the two approaches NASA picked the Hughes design for the mission.

But even though the technical challenges of designing the spacecraft and its instruments had been largely met and overcome, the project encountered the type of unreasonable resistance here on Earth which has marred so many American space endeavors. Political maneuvering almost sabotaged the American mission to Venus as it had plagued so many other U.S. space programs.

Congressional authorization could not be obtained for a mission to start in fiscal 1974. Out of a total national budget of more than $230 billion the relatively meager pittance for the mission to Venus came under attack. As a result program management could not undertake to meet launch dates for the 1976–77 multiprobe mission. Therefore in August 1972 the mission series was changed. Only two launches were planned and these would both be slipped to later years. In February 1973 another important decision was made to delay the Multiprobe launching until 1978 so that both the Multiprobe and the Orbiter would use the same opportunity and arrive at Venus toward the end of 1978, thereby saving on operational costs.

In March 1973, NASA Headquarters decided to use Atlas-Centaur launch vehicles instead of Deltas so that the total cost of the mission could be reduced. As a result of competitive bidding following issue of a request for proposals in June 1973, Hughes Aircraft Company was selected in February 1974 for negotiation of a cost-plus-award-fee contract for continued design of the system. The proposed cost of design work was $3 million, with an option for final design, development, fabrication, and testing of two flight spacecraft, and launch support at $55 million. A contract was awarded in May 1974. After more negotiations a final award, including hardware, was made to Hughes Aircraft Company in November 1974.

Meanwhile by August 1974 Congress had at long last approved a new start in fiscal 1975 for a mission to explore Venus. By the beginning of calendar 1975 work was well underway. Final contracts for scientific instruments were negotiated by June 1975, and in the following 12 months problems of how the instruments should be integrated into the spacecraft had been studied and mostly resolved, and first tests of the parachute system, needed for descent of the Large Probe into the atmosphere of Venus, had been made at Vandenberg Air Force Base, California.

The program still had to face and overcome major fiscal problems before the spacecraft could be launched. Charles F. Hall, program manager, commented at an interview about the fiscal problems of managing a space program. "It always seems you don't have enough time and you are trying to find ways to do things faster. You are always having trouble with funding. You may have a total amount of funds that is enough for the program but you never seem to have enough for any particular year. So you are always making small perturbations to your plans to work around funding difficulties."

A typical example of this type of problem occurred in June 1975. During the budget hearings for fiscal year 1976, a serious political setback occurred to Pioneer Venus. At a critical point, when about $50 million had already been invested in the program, the House of Representatives voted to cut $48 million from the NASA funds intended for the mission to Venus. It was an unexpected and almost fatal blow.

As stated earlier in this chapter, the idea for the program came originally from the scientific community. A group of eminent planetary scientists had stated it was a mistake for the nation not to have a program to explore Venus. They defined which science experiments were important so that the mission could be designed to achieve those goals. Moreover, they received considerable support from scientists.

When the news of the political blow was published many scientists were extremely disappointed and frustrated. They pointed out that the budget cuts could kill the program. If the launch were delayed to the 1980 opportunity, as would have been necessary if the funds were withheld at that time, the spacecraft would need redesigning because the 1980 opportunity was not as favorable as that in 1978. More launch energy or a lesser payload would have been required. This might have spelled *finis* to the program because as much as $50 million additional money over that originally requested for the program would have been needed for a mission to Venus at the less favorable opportunity, and the political climate was to cut space funding, not increase it for any mission.

However, scientists, the national press, and many organizations rallied around the Pioneer Venus cause. Letters and telegrams were sent to congressmen and speeches were made; the legislators were made aware in no uncertain fashion that cancellation of the Venus program was most unwise and would reflect on their credibility of understanding in an increasingly technological and scientific world. The nation's leading climatologists and meteorologists emphasized how important a study of Venus was to their studies of Earth's environment and its protection. The dangers of climatic changes on Earth were cited and the need to be able to predict them was shown as having an enormous potential payoff by preventing mass starvation from failure of agriculture.

As a result of this lobbying, funds for Pioneer Venus were restored by a Senate subcommittee. The senators reversed the House move to slash all but $9.2 million from the project and virtually kill it. But the project still faced further hurdles. The Senate Appropriations Committee and then the full Senate had to approve it, and if they did there would still be the need for a joint committee to work out a compromise with the House of Representatives. The Senate committee acted on the bill later in July and gave support to the mission, and early the next month the Senate also recognized the importance of the program by giving approval to NASA's requested funding of $57 million for Pioneer Venus during fiscal 1976.

During September the go-ahead finally came. The Senate-House conference committee restored all but $1 million of the funds for the project to send the two Pioneer spacecraft to Venus in 1978. The Earth-based part of the mission was back on course. Scientists and engineers could again concentrate on the difficult task of having the spacecraft and their science instruments ready for when Venus and Earth moved to the right planetary configuration for the launches.

Meanwhile, in June 1975, the Pioneer Venus program made use of the largest structure of its type in the world, the Vehicle Assembly Building at NASA's Kennedy Space Center, Florida, originally built for final assembly of the huge Saturn V boosters used to launch Apollo spacecraft to the Moon. The parachute for the Large Probe was an important part of the Pioneer Venus system, for it was essential to slowing the Large Probe on its descent into the atmosphere of Venus. The Vehicle Assembly Building was used for preliminary testing of the parachute design (figure 4.4).

Later the parachute was tested in drops from high-altitude aircraft and balloons so that the speed of the probe would be close to the conditions to be encountered in the mission just before the probe would descend into the dense, hot, lower atmosphere of Venus. The tests were aimed at confirming the deployment of the probe parachute, separation of the atmospheric entry heat shield and, after 17 minutes of parachute descent, separation of the pressure vessel for it to plunge deeper without the parachute. The fast descent after release of the parachute would be needed on Venus so that the probe could penetrate deeply into the atmosphere before the high temperatures could destroy its instruments.

There were many problems in developing the parachute system. "For a time it almost looked as though we were never going to get a parachute," commented Charles Hall after the mission. He related how a newly designed parachute had been taken to the desert near

Figure 4.4. A parachute had to be developed to slow the Pioneer Venus Large Probe on entry into Venus's atmosphere. Here a version of the parachute is undergoing tests in the Vehicle Assembly Building at Kennedy Space Center, Florida. A spherical descent capsule is suspended beneath the parachute. (Photo NASA)

El Centro, California, for it to be tested in a drop from an F-4 airplane. The parachute was attached to a pointed cylinder which carried high-speed cameras and test instruments. When the airplane was traveling at the right speed and altitude the cylinder would be dropped from it and the parachute deployed, a drogue chute pulling out the main chute.

The day of the test arrived. When the F-4 reached altitude and flight speed for the drop, the cylinder was released and the drogue chute deployed. Observers were appalled to see no trace of the main chute's opening. It had disappeared. Hall described how engineers examined film records and saw that the parachute had started to open—and then disintegrated into shreds. The camera had taken pictures at 200 frames/sec, and the projector could display one frame at a time. Said Hall, "You wouldn't believe it, but on one frame the parachute would be intact and on the next frame there would be nothing. It was not that it was breaking away from the shrouds; the material itself just went into shreds."

Engineers said that the test environment was too severe, so another test was planned to apply a lower dynamic pressure on the parachute. Again the parachute shredded in a split second.

A third try also failed. When engineers inspected the pictures they saw that on one frame just before complete failure, many of the parachute gores were missing. This was suspicious. So one of the parachutes was set up and tested in the 40 by 80 wind tunnel at NASA Ames Research Center. The wind speed was reduced to a level at which an engineer could walk inside and watch the opening at close hand. When the parachute opened and some gores still remained folded, he tried in vain to pull them apart. The chute design effectively held the gores together, and it had to be abandoned. The project moved back to an earlier conical ribbon design.

But time was critical, and some chances had to be taken. The new parachute was manufactured but it had to be drop tested; almost immediately there was no time left to try airplane drops first. In addition to the parachute, the heat shield release mechanism and other parts of the system also had to be tested—all on one drop from a high-altitude balloon.

On the day of the test the balloon gently lifted its load from White Sands Proving Grounds, New Mexico, and carried the test vehicle to 30,000 meters (100,000 ft). Engineers at Ames Research Center anxiously awaited results. Then came the dreaded phone call; the test had been a complete failure. The test vehicle had dropped swiftly from the gondola beneath the balloon as planned. But the parachute did not open, and everything plunged toward Earth at high speed, and dug a crater in the desert floor. The project was in deep trouble. Pioneer Venus did not have a parachute.

Quickly photographs were rushed to Ames Research Center for study, and recovered parts of the test vehicle were also inspected. The cause of the failure seemed clear; all over the test vehicle were structural breakages which had occurred before the impact. Not only had the parachute failed but the whole system appeared to have been stressed incorrectly.

Then engineers studied the photographs more thoroughly. They found that the test vehicle had tumbled end over end for part of the way down. Finally it stabilized, but was traveling tail first instead of nose first. The result was that when the parachute deployed, it came out of its canister at an awkward angle and this exerted an enormous stress on the structure before the chute broke away from the body because of the wrong attitude.

The question next to be answered was why the test vehicle had tumbled after it had been released from the container in which it had been carried beneath the balloon. Further investigation produced the answer. Just before the balloon was due to ascend from the ground a test engineer decided there was a chance that as the gondola climbed to 30,000 meters, the temperature would drop to a level at which equipment in the Large Probe might operate unreliably. So he taped a protective blanket of ½-inch fibrous padding beneath the box to insulate it from the cold of the high atmosphere. When the probe was released an edge of it snagged on the blanket and the probe tumbled on its fall. It was not the parachute that was at fault.

Another test vehicle was built and this time the test was successful. At last, and only just in time, Pioneer Venus had a working parachute for its Large Probe.

However, there were new problems still to be tackled. During a thermal vacuum test of the probe spacecraft only seven months before the launch date, batteries within the probe failed. This seemed disastrous. However, the batteries themselves had fortunately not caused the failure. The fault lay in the conditions of the test, by which the spacecraft had been spun on an axis aligned horizontally. This orientation sloshed electrolyte within the batteries and caused the cells to fail. Such sloshing would not occur during an actual mission. All was well.

Making cable connections within the restricted space inside the probes also gave the engineers headaches. All spacecraft have problems with cable harnesses and Pioneer Venus presented even more difficulties than usual. This was because the probes had to be assembled and taken apart several times before launch. The standard procedure for each spacecraft after design and development was complete, was to assemble and test it, then take it apart again so that the principal investigators could have their instruments for final calibrations in their laboratories. Then the instruments were replaced in the spacecraft before the whole was shipped to the launch area for mating with the launch vehicle. And at the launch site there is often need to service a spacecraft. Once again the problems were solved by the engineers responsible for the spacecraft design.

Another major challenge was how to seal the probes. On the way to Venus the pressure was directed outward, so that leakages into the vacuum of space had to be avoided. After entering the atmosphere the probe would encounter very high external pressures and then would have to withstand inward leaks. Two types of seals had to be used for these opposing conditions; O-rings to seal against the vacuum of space, and flat graphoil seals to resist the pressure of Venus's atmosphere. The system worked so well that when one probe continued sending engineering data for a short while from the surface of Venus the data showed no evidence of any leakage.

There were, however, many development problems encountered in making a suitable seal for the Large Probe's relatively big diamond window, which was needed for transparency to infrared radiation. An early decision was not to braze the window to seal it to the shell of the pressure vessel. As the program continued the window sealing presented a very difficult fabrication problem and there were many disappointments in developing an effective diamond window seal. Eventually a mechanical flat seal proved satisfactory. However, it became such a pacing item that before launch the flight diamond window with the full assembly could not be tested with the instrument actually in the spacecraft. So the final test had to take place in the atmosphere of Venus.

At the beginning of the program the Thor-Delta was planned as a candidate launch vehicle for the Pioneer mission. Later the Thor-Delta was suggested as being the launch vehicle for the Orbiter while an Atlas-Centaur would be needed for the Multiprobe because of its greater weight. But maintaining interfaces between different vehicles and spacecraft could easily have cancelled any saving from having two launch vehicle types for the mission. In fact, the Thor-Delta could not have been used for the mission because during the development of the spacecraft the Orbiter became too heavy for that launch vehicle. It was fortunate that Pioneer project management were able to convince everyone concerned that the two launch vehicles had each to be an Atlas-Centaur (figure 4.5), which was NASA's standard launch vehicle for payloads of intermediate weight.

The 40-meter (131-ft) high launch vehicle for each spacecraft consisted of an Atlas SLV-3D booster with a high-energy Centaur D-1A second stage. The spacecraft was enclosed in a fiberglass nose fairing to protect

Figure 4.5. Atlas-Centaur, NASA's standard launch vehicle for payloads of intermediate weight, was picked for the Pioneer Venus mission. (Photo NASA)

it as the launch vehicle accelerated through Earth's atmosphere into the vacuum of space where the fairing could be jettisoned.

Pioneer Venus control and spacecraft operations were located at the Pioneer Missions Operations Center (PMOC), at Ames Research Center, Mountain View, California. Operations were complicated by the continued operation of previously launched Pioneer spacecraft; Pioneers 6 and 9 circling the Sun, Pioneer 10 heading out of the Solar System through previously unexplored regions of space, and Pioneer 11 on its way to the first rendezvous of a spacecraft with Saturn.

Under operational direction of the flight director for the Venus mission, all command information to the spacecraft originated from the Mission Operations Center. The PMOC also received all telemetry data required to control the mission.

All six spacecraft of the Pioneer Venus mission—four Probes, the Bus, and the Orbiter—were tracked by the Deep Space Network's global system of large antennas. The largest of these antennas was essential for critical phases of the mission such as reorientation of the spacecraft, velocity corrections, orbit insertion, entry of the four probes into Venus's atmosphere, and radio occultation experiments.

The Deep Space Network's facilities are located at approximately 120° intervals around the Earth (figure 4.6). As the Orbiter and the Multiprobe set at one station due to the rotation of the Earth, they rose at the next. During the critical two-hour period of atmospheric entry by the Bus and flights down to the surface by the four probes, the 64-m (250-ft) antennas at Goldstone and Canberra were both needed to gather the data arriving simultaneously from all the spacecraft.

Incoming telemetry data from the spacecraft were formatted at the Deep Space Network stations and transmitted to the computers at Ames Research Center. These computers looked for abnormal data from the spacecraft and its instruments. Should any abnormalities be found, information would then be provided for analysis by specialists experienced in all details of the spacecraft, the experiments, and the ground system. This computer watchfulness ensured that all the spacecraft were always controlled correctly to get the best science

Figure 4.6. The communications network for the Pioneer Venus mission included the big antennas of the Deep Space Network at Goldstone, California, and Canberra, Australia, and other supporting ground stations as shown in this map. (NASA/ARC)

results. Outgoing commands were verified by the computers at Ames Research Center and sent to the Deep Space Network stations, where they were again verified by computer before being transmitted to the spacecraft. Navigation data and trajectory computations for the spacecraft were furnished by the Jet Propulsion Laboratory.

The Deep Space Network had to be modified to meet the special needs of Pioneer Venus. Extra receivers were added to handle five different data streams. Wideband recorders accommodated large frequency drifts caused by changes in velocity of the probes as they plunged into the atmosphere. To protect data during this critical period, the Deep Space Network also used equipment to tune the receivers automatically.

Incoming telemetry of engineering and science data from the spacecraft was transmitted over the NASA Communications System (NASCOM) to the Pioneer Mission Computing Center (PMCC) for processing. Information was quickly displayed to show the state of all the spacecraft and their experiments, and experimental data records were supplied to each of the principal investigators for distribution to their team members.

In February 1978 preshipment reviews of the Orbiter took place at Hughes Aircraft Company, in El Segundo, California. Verbal and written reports were given to an audience of engineers and scientists on all aspects of the spacecraft to confirm that specifications were met and the machine was ready for the interplanetary mission. Then the main body of the spacecraft and the high-gain antenna were shipped separately to the launch site at Kennedy Space Center, Florida. The first task on arrival in Florida was to mate the antenna with the spacecraft. Then the Orbiter was put through an extensive series of tests to make sure that all its subsystems and scientific instruments operated as specified.

Next, ordnance was installed that would not be harmful to the spacecraft or test personnel should it inadvertently be fired. The spacecraft was transferred to Building SAFE-2 where the rest of the ordnance and 32 kg (70 lb) of hydrazine propellant—for trajectory correction and orientation maneuvers—was loaded. Then the spacecraft was mated with a launch vehicle adapter and transferred to launch pad 36, where it was mated to the waiting Atlas-Centaur.

On the launch vehicle the spacecraft underwent more tests to verify that there had been no deterioration of its systems. Radio frequency interference tests verified that the radar needed to track the vehicle during launch would not interfere with the spacecraft nor affect the data being transmitted from it during this period.

Next, several practice countdowns were made, followed by a final test with the Deep Space Network to make sure that expected test signals from the spacecraft were received at the required signal strengths. After some ten days of various tests the go for launch was given and the final countdown began. There were no holds. The Atlas-Centaur thundered into the Florida skies on May 20, 1978, sending the Orbiter on its way to Venus (figure 4.7).

Meanwhile during April, the Multiprobe had completed its pre-shipment review in El Segundo and the Large Probe was shipped separately from the Bus and the three Small Probes. On arrival at Kennedy Space Center, the Small Probes were removed from the Bus

Figure 4.7. Pioneer Venus Orbiter was launched from Florida on May 20, 1978. (Photo NASA)

and each was thoroughly checked, as was the Large Probe. A unique feature of this checkout was that the flight batteries had to be kept under very strict thermal control. They could not be onboard the probe spacecraft for testing because they could not be put through charge/discharge cycles. Other batteries were used for the tests, and the flight batteries were the last items to be installed on the probe spacecraft at the end of checkout.

Then pressure vessel sealing was checked by sending the probes to Martin Orlando, where they were tested in a thermal vacuum chamber. The gas inside each probe contained a small amount of helium. A failure of the seals would be revealed by the presence of helium in samples taken periodically from the vacuum chamber. None was found. All probes passed the test and were returned to the Space Center to be placed on the Bus. Next the pyrotechnics were installed—explosive bolts for the Large Probe and bolt cutters for each of the Small Probes—to release the probes from the Bus before arriving at Venus. Hydrazine for maneuvers was also loaded into the Bus.

After the Multiprobe was transferred to the launch pad it was mated with its Atlas-Centaur for final checkout there. Awaiting launch the probes were warm; close to the ambient temperature in Florida. When they would reach Venus's atmosphere they were expected to be very cold. So as not to exceed temperature limits established for the equipment carried within the heavily insulated probes, tests that generated heat within a probe had to be extremely brief. The radio frequency transmitter on each probe was turned on for a short time only to verify that it did emit a signal.

When the Multiprobe entered its countdown all went well until close to the scheduled launch date of August 6. Then, as the liquid helium was loaded into the Centaur, a major problem surfaced; there was insufficient helium in the tank of the truck. The countdown had to stop. It began again on August 7 at 6:30 P.M. EDT. This time all went smoothly and the Multiprobe was successfully sent on its way to Venus in a spectacular night launch on August 8, 1978 (figure 4.8).

The Orbiter spacecraft provided a spin-stabilized platform for the twelve scientific instruments. The spacecraft itself consisted of six basic assemblies: a despun antenna system, a bearing and power transfer assembly and its support structure, an equipment shelf, a solar array, an orbit insertion motor and its case, and a thrust tube. These are identified in figure 4.9).

The spacecraft's main body was a flat cylinder 2.5 meters (8.3 ft) in diameter and 1.2 meters (4 ft) high, in the forward end of which a circular equipment shelf carried all the scientific instruments and electronics. The shelf was mounted on the end of a thrust tube to connect the spacecraft to the launch vehicle. Fifteen thermal louvers controlled heat radiation from the equipment compartment. A cylindrical array of solar cells covered the circumference of the flat cylinder of the bus. A solid-propellant rocket motor provided 18,000 Newtons (4000 lb) of thrust to decelerate the spacecraft into orbit around Venus.

On top of the spacecraft a 1.09-m (43-in) diameter,

Figure 4.8. A few months after the Orbiter, the Pioneer Venus Multiprobe was dispatched to Venus on August 8, 1978, in a spectacular night launching. (Photo NASA)

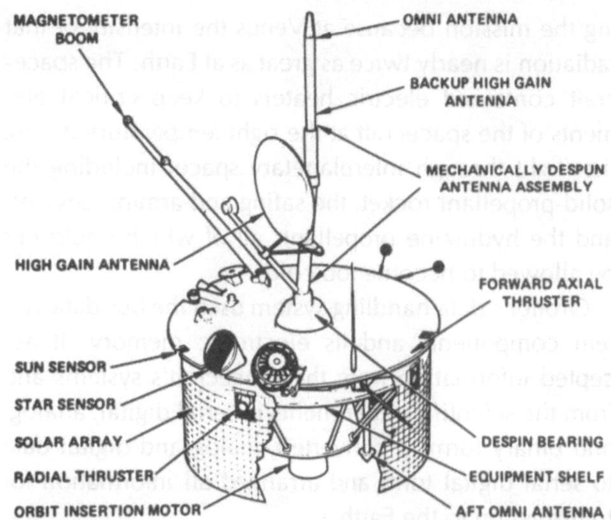

Figure 4.9. Diagram of the Pioneer Venus Orbiter spacecraft showing its main components. (NASA/ARC)

high-gain, parabolic dish antenna, which operated at S- and X-bands, was mounted on a mast so that its line-of-sight cleared all other equipment outside the spacecraft. The design allowed the antenna to be mechanically directed to Earth from the spinning spacecraft. Including the antenna mast the Orbiter was almost 4.5 m (15 ft) high and it weighed 553 kg (1219 lb) including 45 kg (100 lb) of scientific instruments and 179 kg (395 lb) of rocket propellant.

A maneuvering system for the basic Bus controlled its rate of spin, made course and orbit corrections, and maintained the orientation of the spin axis, which was usually kept perpendicular to the plane of Earth's orbit. Beneath the equipment compartment and attached to the thrust tube were two propellant tanks to store 32 kg (70 lb) of hydrazine for use in three axial and four radial thrusters.

Two axial thrusters were aligned with the axis of spin at top and bottom of the Bus cylinder diagonally opposite from one another and pointed in opposite directions. To turn the spin axis, the thrusters were fired in pulses. To speed up or slow down the Bus along the direction of its spin axis, only one thruster need be fired in pulses at two points 180° apart in the rotation of the Bus. The top or the bottom thruster was chosen to do this depending upon the direction required for the change in velocity. The third axial thruster, located at the bottom of the thrust cylinder, permitted continuous firing of two bottom thrusters to make moves in an axial direction when needed to change the orbit of the spacecraft.

The four radial thrusters were arranged in two pairs, pointing in opposite directions and mounted in a plane perpendicular to the spin axis—a plane which passed through the spacecraft's center of gravity. These thrusters were used to change the Orbiter's velocity in a direction perpendicular to the spin axis, or to control the spin rate. Firing two thrusters 180° apart slowed the spin rate. Firing the other two increased it. A star sensor and Sun sensors provided attitude references to control the spacecraft.

The thermal design isolated the equipment from extremes of temperature arising from solar radiation during the mission because at Venus the intensity of that radiation is nearly twice as great as at Earth. The spacecraft contained electric heaters to keep critical elements of the spacecraft at the right temperature during the flight through interplanetary space, including the solid-propellant rocket, the safing and arming devices, and the hydrazine propellant; all of which could not be allowed to become too cold.

Orbiter's data handling system used the bus data system components and its electronic memory. It accepted information from the spacecraft's systems and from the scientific experiments in serial digital, analog, and binary form. It converted analog and digital data to serial digital form and arranged all information for transmission to the Earth.

An important part of the Orbiter that was not included in the basic bus used for Multiprobe was the despun high-gain parabolic antenna. At S-band, this antenna directed a tight beam toward the Earth throughout the mission and concentrated the signal 316 times. The antenna was designed to return data at high data rates over the greatest distance that the Orbiter would be from Earth—i.e., near superior conjunctions. The multiprobe bus did not require such capability.

The antenna dish, a dipole antenna, and a forward omnidirectional antenna were all mounted on a mast which projected 2.9 m (9.8 ft) along the spin axis from the top of the Orbiter's basic bus cylinder. This group of antennas was despun relative to the spinning spacecraft. A control system had redundant electronics to control the despin mechanism and to drive either one of two despin motors. The dipole radiated in a flat pattern perpendicular to the spin axis to provide a backup if the high-gain antenna could not be pointed toward Earth due to failure of the despin mechanism.

An aft omnidirectional antenna coupled with the forward omni provided low-gain radiation in all directions so that at any orientation of the spacecraft it could receive commands from and communicate at low bitrates with Earth.

To probe more effectively by radio into the atmosphere at occultation the Orbiter carried a 750 milliwatt X-band transmitter. The signal frequency of this transmitter was maintained proportional to that of the main S-band transmitter. Both X- and S-band signals were transmitted by the directional antenna which could be moved 15 degrees from the Earth-line as the Orbiter passed behind Venus. This aiming of the antenna permitted the two radio beams to dip deeply into the atmosphere of Venus and still reach Earth in spite of refraction by Venus's atmosphere. In this way radio occultation data were obtained deeper into the atmosphere at two radio frequencies.

Commands from Earth were received at any orientation of the spacecraft through two identical S-band transponders connected to the omnidirectional antennas. Each transponder received the radio signal from Earth and tuned the transmitter so that the frequency of the outgoing radio signals from the spacecraft bore a constant ratio to the frequency of the incoming signals. As a result the Doppler shift in the radio frequency arising from the motion of the spacecraft relative to the Earth could be measured precisely both on the outgoing and incoming radio signals and thereby allow the velocity of the spacecraft to be determined to within 3 m per hour.

The receiver of each transponder responded to only certain frequencies. If no command was received from Earth in a period of 36 hours, the receiver that was off was automatically switched on to ensure that if one receiver should fail the other would take over within 36 hours without a command from Earth.

The Multiprobe spacecraft (figure 4.10) consisted of the basic Bus, a Large Probe, and three identical Small Probes. It weighed 875 kg (1930 lb), including 32 kg (70 lb) of hydrazine for correcting the trajectory and orienting the spacecraft's spin axis. The weights of individual spacecraft were: Bus, 290 kg (641 lb); Large Probe, 315 kg (695 lb); and Small Probes, each 90 kg (198 lb).

The basic Bus design was similar to that of the Orbiter and it used a number of subsystems common to the Orbiter. It had five subassemblies; a support structure for the Large Probe, a support structure for the Small Probes, an equipment shelf, a peripheral solar array, and a central thrust tube. The spacecraft's diameter was 2.5 m (8.3 ft). From the bottom of the Bus to the tip of the Large Probe mounted on it measured 2.9 m (9.5 ft).

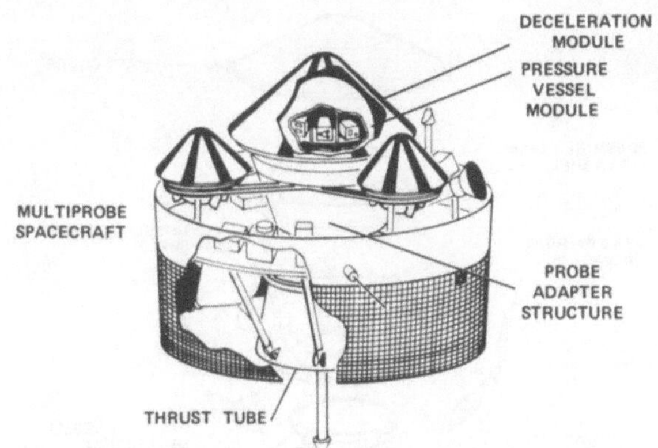

Figure 4.10. Diagram of the Multiprobe spacecraft showing its main components. During flight from Earth to Venus the four probes were carried on the Multiprobe Bus to be released just before arrival at the target planet. (NASA/ARC)

During the flight to Venus the four probes were carried on a large inverted cone structure secured by three equally spaced circular clamps surrounding the cone. These attachment structures were bolted to the thrust tube of the Bus which formed the structural link to the launch vehicle. The Large Probe was centered on the spin axis of the Bus and was launched from it toward Venus by springs released by a pyrotechnic device. The ring support clamps that attached the Small Probes were hinged. When explosive nuts were fired, these clamps opened to launch the Small Probes which spun from the Bus tangentially because of its rotation at 48 rpm.

The Multiprobe's forward omnidirectional antenna extended above the top of the Bus cylinder. An aft omni antenna extended below it. Both had hemispherical radiation patterns. A medium-gain horn antenna was attached to the instrument shelf and radiated aft of the spacecraft. It was used during critical maneuvers when the aft of the spacecraft pointed toward Earth at the time the probes separated from the Bus.

The instrument-equipment compartment, as in the Orbiter, carried the scientific experiments and electronics for the spacecraft's subsystems. The solar array provided electrical power from solar radiation. There were batteries and a power distribution system, Sun and star sensors, propellant storage tanks and thrusters for maneuvering and stabilization. The Bus also carried radio transmitters and receivers, data processors, and a command and data handling system.

The thermal design was essentially the same as that of the Orbiter. In addition, however, the Bus required protective surfaces in the vicinity of the Small Probes to keep them at the required temperature during the cruise and to protect the Bus itself from heating after the Probes had separated from it.

Except for the positioning of the high-gain antenna on the Orbiter, the orientation controls for the Multiprobe were the same as those for the Orbiter. The propulsion system was identical to that of the Orbiter except the Multiprobe had only one aft axial thruster.

The data handling system for the Multiprobe was similar to that of the Orbiter, but with data formats organized to meet the special requirements of the Multiprobe's mission. Before separation of the probes from the Bus, the Multiprobe handled data for the Bus and all the probes. The data system of the Multiprobe accepted engineering and missions operations information from the four probes as well as from the Multiprobe Bus itself and from the experiments carried on the Multiprobe Bus. It converted analog data to serial digital binary form and arranged all the information for transmission to Earth. After separation the probes used their own data systems.

The communications subsystem of the Multiprobe received and transmitted radio communications from and to Earth. There were two redundant receiving channels and hemispherical onmidirectional antennas, so that together they provided coverage in every direction around the spacecraft. Transmissions from spacecraft to Earth could be assigned by command to either of the hemispherical omnidirectional antennas or to the horn antenna. The spacecraft transmitted at a frequency that was related to the uplink frequency. When there was no signal being received from Earth, the transmission frequency from the spacecraft was governed by a crystal oscillator within the spacecraft.

Designing the probes to survive in the atmosphere of Venus presented a great technical challenge. They had to contend with high pressure in the lower regions of the atmosphere—about 100 times that of the Earth's atmospheric pressure at sea level—a high temperature of about 480° C (900° F) at the surface, and the corrosive constituents of the atmosphere, such as sulfuric acid. In addition the probes must hurtle into the atmosphere at about 41,600 kph (26,000 mph), which is 43 times the speed of a typical commercial jet.

The main component of each probe was a spherical vessel, machined from titanium and sealed against the vacuum of space and the high pressure within the atmosphere of Venus. Inside this vessel were scientific instruments and systems to operate the spacecraft, to supply it with power, to accept commands from Earth, to gather data, and to transmit it to Earth.

Each pressure vessel was enclosed in an aeroshell and an aft shield. The conical, blunt-tipped aeroshell was an aluminum structure with stiffening rings. It protected the probe from the heat generated during the plunge into the atmosphere and it stabilized the flight of the probe. All instruments within the pressure vessels of the probes required either observations or direct sampling of the atmosphere of Venus. Providing access to the atmosphere while maintaining efficient seals was a major design problem. Special windows of diamond and sapphire were used to admit light or heat at wavelengths required for several of the science experiments.

The Large Probe (figure 4.11) weighed about 316 kg (695 lb) and was about 1.5 m (5 ft) in diameter. It consisted of a forward aeroshell-heat shield, a pressure vessel, and an aft cover. For entry into Venus's atmosphere the Large Probe was protected from overheating by its ablative heat shield of carbon phenolic which was bonded to and covered the outer surface of the forward-facing aeroshell. All other surfaces of the aeroshell and the aft cover were coated with a heat-resisting, low-density, elastomeric material.

The pressure vessel was machined from titanium to reduce weight but achieve high strength at elevated temperatures. It was 73.2 cm (28.8 in) in diameter and made in three flanged pieces; an aft hemisphere, a flat ring section, and a forward cap. The pieces bolted together with combination seals between the flanges. An O-ring prevented leakage of the 102 kiloPascals (kPa) [15 pounds per square inch absolute (psia)] nitrogen atmosphere of the probe in space, and graphoil flat gaskets prevented later inward leakage of the hot Venus atmosphere. A pressure bottle, mounted on the forward shelf of the Large Probe, could be opened by command to increase the probe's internal pressure by 6 psi. Inside the pressure vessel two parallel shelves made of beryllium served as supports and as heat absorbers for the instruments and spacecraft systems mounted on them.

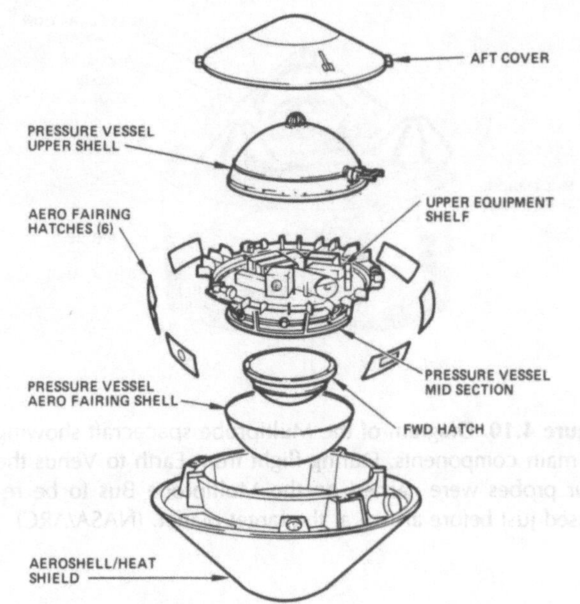

Figure 4.11. Exploded diagram of the Large Probe showing its various components. (NASA/ARC)

Four science instruments (figure 4.12) used nine observation windows through four of the pressure vessel penetrations mentioned earlier. Eight windows were of sapphire and one of diamond. A solar flux radiometer had five windows through one hull penetration; a nephelometer had two windows, and infrared and cloud particle instruments had one window each. Three vessel penetrations were inlets for direct atmospheric sampling by a mass spectrometer, a gas chromatograph, and an atmospheric structure equipment. At the aft pole of the pressure vessel an antenna provided a hemispherical radiation pattern to communicate with Earth when the probe separated from its Bus. Two arms extended 10 cm (4 in) on one side of the pressure vessel to hold a reflecting prism used in the cloud particle observations. On the opposite side of the pressure ves-

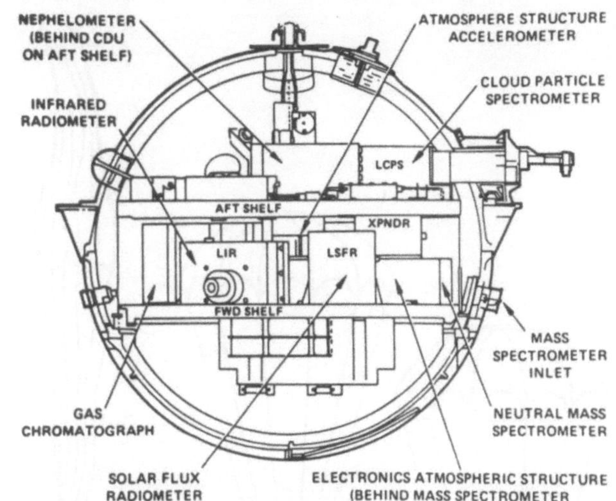

Figure 4.12. Science instruments carried by the Pioneer Venus Large Probe. (NASA/ARC)

sel a single arm carried a temperature sensor. Three parachute-shroud towers were mounted above aerodynamic drag plates.

The fiberglass honeycomb aft cover had a teflon flat section transparent to radio waves. It protected the aft hemisphere of the pressure vessel during entry into Venus's atmosphere. Spin vanes maintained the probe's stability during descent.

Communications with Earth started 22 minutes before atmospheric entry. A peak deceleration of 280 g's occurred soon after entry, then the aft cover was jettisoned. A pilot chute was attached by lines to the aft cover which was freed by an explosive bolt. The pilot chute then pulled the main chute from its compartment within the aeroshell. As soon as the probe was stable, mechanical and electrical ties to the aeroshell were severed by explosive nuts or by cable cutters, and the main chute then pulled the pressure vessel free from the aeroshell (figure 4.13). The heat shield was jettisoned about 67 km (42 mi) above the surface. At about 47 km (30 mi) altitude the parachute was released and the probe fell freely. It took about 55 minutes to reach the surface.

The Large Probe's solid state 40-watt transmitter returned a stream of data directly to Earth at 256 bits per second (bps). Power was provided by a 40 ampere-hour silver-zinc battery. Once the probe separated from the Bus, its internal electronics provided all commands needed to operate it. It contained a cruise timer which was the only part of the spacecraft to operate during the period from separation from the Bus to entry into the atmosphere of Venus. During this interval all other systems within the probe were shut off. An entry sequence programmer was wired to transmit 53 assigned discrete commands into a preset sequence. Commands were initiated by a clock generator or by a gravity switch that sensed deceleration. A temperature switch provided a backup for the timer when the parachute was jettisoned.

The three Small Probes (figure 4.14) were identical. They had no parachute but were slowed by aerodynamic braking. Each Small Probe consisted of a forward heat shield, a pressure vessel, and an afterbody. The heat shield and the afterbody remained attached to the pressure vessel all the way to the surface. Each probe was 0.8 m (30 in) in diameter and weighed 90 kg (200 lb).

The pressure vessel of machined titanium nested within the aeroshell and was permanently attached to it. The seals between the two halves of the pressure vessel, like these on the Large Probe, consisted of O-rings to maintain internal pressure in space and graphoil flat gaskets to keep out the hot Venus atmosphere. Each Small Probe was filled with xenon at a pressure of approximately 102 kPa (15 psia) to reduce the flow of heat from the pressure vessel walls to the instruments and spacecraft systems. The titanium aeroshell had the same blunt cone design as the Large Probe, with a bonded carbon phenolic ablative coating as a heat shield.

Communications started 22 minutes before entry. Next, about 5 minutes before entry, two weights were cut loose by a pyrotechnic cable cutter. They swung out like yo-yos on 2.4-m (8-ft) cables to reduce the spin rate of each probe from about 48 rpm to 17 rpm before being jettisoned. Aerodynamic forces aligned the probes so that their heat shields could protect them from the heating of entry. All probes hurtled into the atmosphere at about 42,000 kph (26,000 mph). After deceleration of the probe, three doors on the afterbody then opened

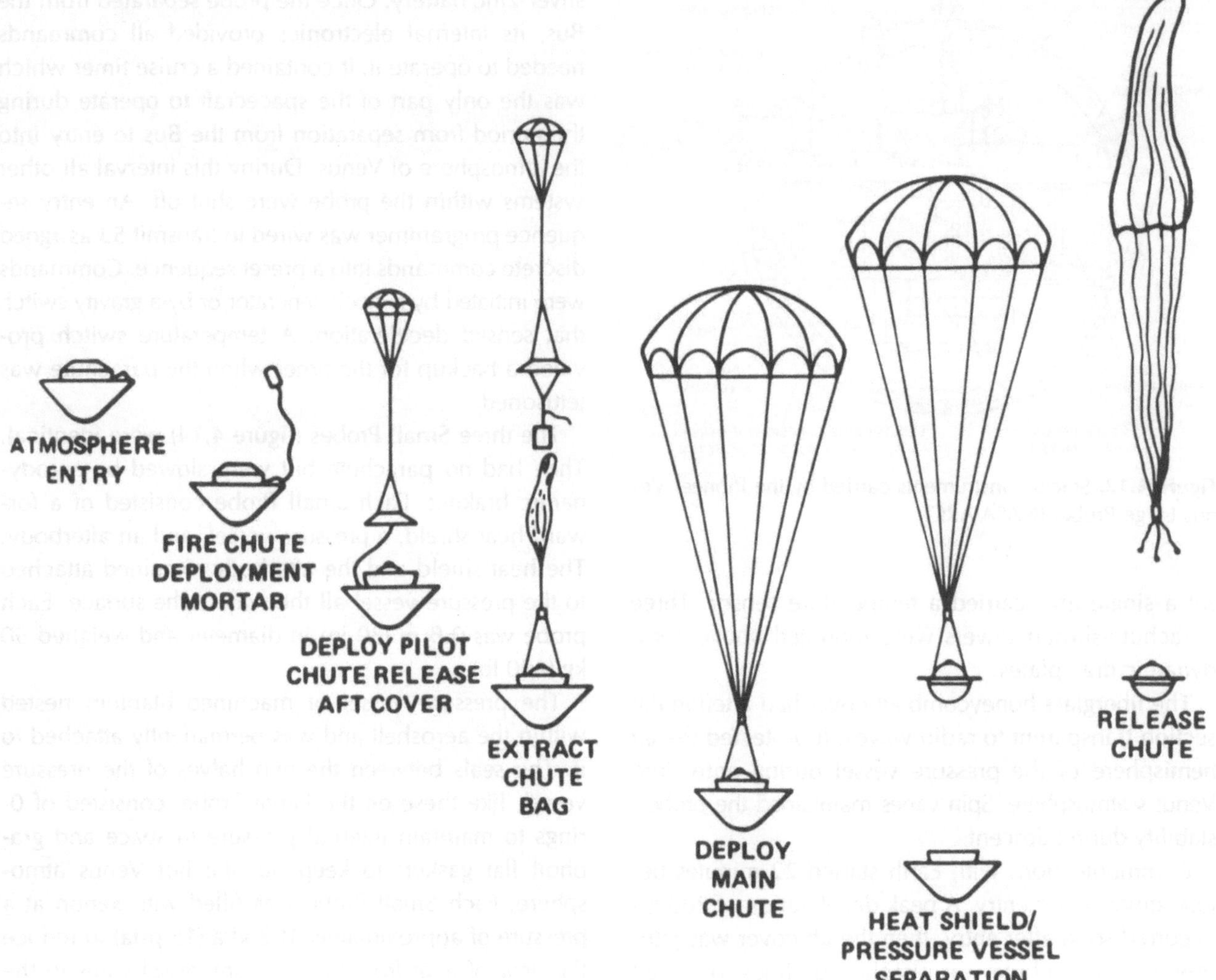

Figure 4.13. Sequence of descent of the Large Probe into the atmosphere of Venus showing deployment and subsequent release of the parachute. (NASA/ARC)

at an altitude of about 70 km (44 mi) to provide access to the atmosphere by instruments carried by the probes. Each Small Probe fell freely to the surface, taking about 53–55 minutes to reach it.

Communications for each Small Probe consisted of a solid state transmitter and a hemispherical-coverage antenna, the same as for the Large Probe. The antenna was mounted at the aft of the pressure vessel and radiated through a teflon window. The transmitter had one 10-W amplifier; that is, one-quarter the power of the transmitter of the Large Probe. Unlike the Large Probe, the Small Probe did not carry a receiver for two-way Doppler tracking. Instead, tracking was by use of a stable oscillator, carried by each probe, to provide the reference frequency for the Doppler measurements. Power came from an 11 ampere-hour, silver-zinc battery.

The command system was identical to that on the Large Probe. There was no way to send commands to the probes. All commands originated onboard the probes

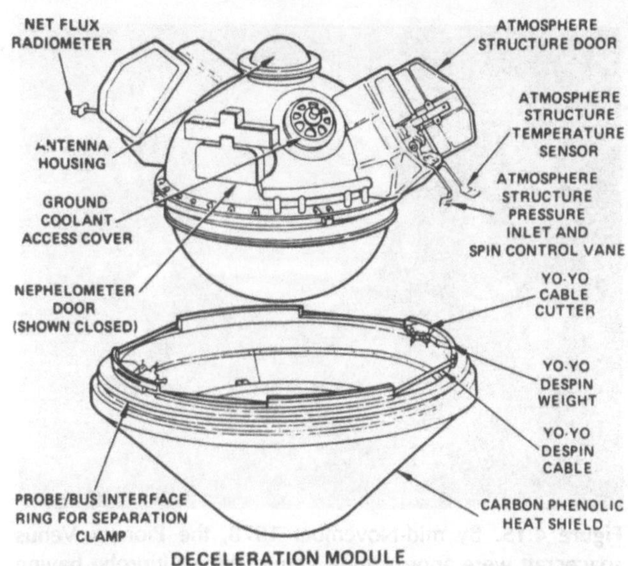

Figure 4.14. Exploded diagram of one of the three identical Small Probes showing the major components. (NASA/ARC)

from such devices as timers, programmers, and gravity switches. Control was maintained by the coast timer. It started an onboard programmer which issued 41 commands in a fixed sequence from the start of the programmer until impact of each probe with the surface of the planet.

There were exciting incidents during the long flight of the Orbiter and Multiprobe spacecraft through interplanetary space. One occurred at the time of the Orbiter's first significant maneuver after leaving Earth. Soon after the launch on May 20, 1978, the spacecraft's long magnetometer boom had been deployed, and the dish antenna had been despun so that it could face toward Earth from the spinning spacecraft. The Orbiter and several of its scientific instruments had been checked, and telemetry indicated that everything was all right. The spin-scan imaging system had been tested by obtaining several pictures of the Earth showing our planet illuminated as a thin crescent.

Controllers commanded a first mid-course correction on June 1, to change the velocity of the spacecraft by 12 kph (7.8 mph) and aim the spacecraft closer to the point near Venus where the Orbiter would fire its rocket motor and achieve an orbit around that planet.

The maneuver was not completed. The cause turned out to be trivial, but it provided the first of many operational lessons learned by mission controllers during the interplanetary voyage. The roll reference system had been designed with an automatic shutoff as a safety feature. A servomechanism followed changes about the roll axis at a restricted rate. If the spacecraft rolled too quickly the servomechanism lost synchronization, and if this occurred during a maneuver the protective design halted the maneuver. It turned out that part of the structure of the spacecraft deflected the gas from the thrusters and rotated the spacecraft. This changed the roll rate sufficiently to overdrive the servomechanism and cause the maneuver to abort automatically. Once they had identified the cause controllers were able to issue commands to disable the automatic cutoff circuit when it was safe to do so.

The first maneuver was then successfully completed. But it took eight hours and thrusts in two directions. This changed the scheduled arrival at Venus so that the Orbiter would reach the orbital injection point some 348 km (216 mi) above the planet's northern hemisphere to enter an elliptical orbit tilted 75° to the equator. Closest point to Venus on the orbit (periapsis) would be 241 km (180 mi), and the greatest distance (apoapsis), would be 66,000 km (41,000 mi). The spacecraft would arrive at Venus at 8:00 A.M. PST on December 4, 1978.

Some significant new science was done on the way to Venus. For example, by early June, Pioneer Orbiter detected an extremely powerful burst of gamma radiation. Gamma ray bursts, first identified in 1973, possess enormous energies and occur about once a month, seemingly from random points in the Galaxy or even from beyond. Two other spacecraft—Vela, a Department of Energy satellite circling Earth, and Helios B, a NASA-European spacecraft orbiting the Sun—also observe the gamma ray burst. In all, the Orbiter recorded six gamma ray bursts, two of which were equivalent to the most powerful so far recorded. Triangulation of several observations as Pioneer moved farther from Earth enabled scientists to locate the origin of the bursts. On March 5, 1979, during the extended orbital mission, Orbiter's instrument recorded a burst of gamma rays which, coupled with observations from other spacecraft, identified the burst as originating from a supernova remnant in the Large Magellanic Cloud.

The Multiprobe spacecraft successfully completed its first course change on August 16. Without a course adjustment the Multiprobe would have missed Venus by about 14,000 km (9000 mi). The course correction took about one day, during which a series of timed rocket thrusts in two directions increased the speed of the spacecraft by 2.25 meters/sec (about 5 mph).

Separating the four probes from the Bus provided two of the most crucial and exciting operations of the entire Venus mission. An error could have made a probe miss its target or fail on entry. The Large Probe was scheduled for release on November 15. Even more critical was the scheduled release on November 19 of the three Small Probes simultaneously and within a few thousandths of a second of a preselected time.

Precisely calculated numbers were placed by command in timers aboard each probe. Each number represented the millions of seconds between the release of a probe and the time when its various systems would operate to begin its entry mission. The probes could be released over a period of three or four days but once a time had been selected the timers had to be set (by command from Earth) for precisely that time. They could not be changed later. If the timers were set short the probe would start up too soon and use all its battery power before reaching the surface. If set too long a lot of data would be missed in the high atmosphere.

By mid-November 1978 both Pioneer Venus spacecraft were converging on their target (figure 4.15). Venus had passed inferior conjunction and a close approach to Earth and was emerging from the Sun's glare, rising as a morning star just before the Sun. The Multiprobe had caught up with the Orbiter and was being readied for separation of the first of its four probes. On December 5 the Orbiter would be placed in orbit around Venus, and five days later the probes would make their hour-long descent through Venus's atmosphere.

The spin axis of the Bus was kept perpendicular to the Earth's orbital plane on the journey from Earth to Venus. On November 9, the spin axis of the Multiprobe was moved through 90° so that the medium-gain,

Figure 4.15. By mid-November 1978, the Pioneer Venus spacecraft were approaching Venus, the Multiprobe having caught up with the Orbiter during the interplanetary cruise. The probes were released from the bus targeted for entry at various regions of the planet. (NASA/ARC)

aft horn antenna of the spacecraft could be used to communicate with Earth. The omnidirectional antenna was no longer suitable to communicate with Earth for checking the probes before their release. Before the Large Probe could be separated from the Multiprobe Bus on November 15, the Bus had to be oriented so that the probe would travel in the right direction.

There were some fears that moving the spin axis would turn the solar panels away from the Sun and they would not be able to produce enough power to maintain the Bus battery sufficiently charged. So when the spacecraft had been reoriented, the attitude and the spin rate had to be measured and adjusted if necessary, and the probes released within a period that would not deplete the battery of the Bus. At about 13 million km (8 million mi) from Venus the spin axis of the Bus was aligned so that the large Probe's entry trajectory would permit the probe to enter the atmosphere with its heat shield correctly aligned relative to its entry flight path.

However, when the spin axis had been changed, tracking data received by the Deep Space Network were startling. The data did not seem right and there was a question as to which direction the Bus was pointing toward. A decision had to be made on whether or not a compensating maneuver was needed.

A big problem in determining orbits is to measure the north-south component of velocity relative to the Earth. This measurement is made by comparing the difference in Doppler frequency at a station in the Earth's Northern Hemisphere with one in the Southern Hemisphere. When many maneuvers have to be made, as was true for the Multiprobe with its requirements of reorienting the antenna and targeting the Large Probe and then reorienting and adjusting the center of gravity location for the release of the Small Probes, very complicated records were necessary to track what was done to the spacecraft's velocity vector and how the spacecraft approached the planet. The long trajectory tracking history during the voyage from Earth was compromised by the preseparation maneuvering. Navigators were concerned that the orientation had not been measured precisely enough, or that the plume of the thrusters had bounced off structure and given a sideways kick to the spacecraft.

In tracking a spacecraft navigators build the trajectory to a current position based on the spacecraft's previous positions. A spacecraft travels along a trajectory calculated from the laws of celestial mechanics. But it is observed from a tracking station on a rotating Earth which itself is traveling in orbit around the Sun and is perturbed by the Moon. The trajectory is calculated and then the observed trajectory is compared with that calculated. Next, the calculated trajectory is refined until the two correspond. Perturbations not included in the calculations begin to show up after they have influenced the trajectory for some time. Navigators measure frequency shifts resulting from the Doppler effect, and Doppler residuals are the differences between the Doppler shift expected from the calculated trajectory and that observed in the signal from the spacecraft. The residuals are continually determined, evaluated, and used to update the calculated trajectory. Accuracies of a fraction of a millimeter per second are achieved.

Before any maneuver is made the anticipated Doppler effect is calculated. If, after the maneuver has been completed, there is a difference between the observed and the expected Doppler residual it is attributed to either of two things; the maneuver was not made in the direction planned, or the performance of the thruster was not as expected.

Judgment is involved in deciding which is which. If the orientation is known, then the residual must be due to the thrusters. And that is especially so if the alignment of the spin axis is, say, 60 degrees from the direction in which the Doppler effects are being observed. It is only when the direction is perpendicular to the line of sight from Earth that there is an unknown situation. So navigators try to do all maneuvers in an alignment which is turned somewhat toward Earth.

Project management thought that a propellant leak might have pushed the spacecraft from its commanded orientation. They needed to know for sure before the probe could be separated at 6:00 P.M. on November 15. There were so many unknowns at that time that Charles Hall, the project manager, decided not to separate the probe until the problem was understood. It took an all-night session of about 12 hours to obtain this understanding. The answer appeared to be solar radiation acting on the spacecraft and pushing it slightly off course.

Because the Large Probe could enter the atmosphere of Venus anywhere over a fairly large area, the aiming point was not critical, but the timer setting was. Timers within the probes had to be set before each probe could be released from the Bus. The number to be set in each timer was very large but had to be correct to one second. This time governed when the probes would wake into life ready for their descent into the atmosphere of Venus. Time was calculated from the instant when each timer would be started by command from Earth, so the time it took the command to travel from Earth had to be included. Hall decided not to attempt another correcting maneuver but to select a setting of the timer that would be the midpoint of extremes for a successful mission. Nevertheless, the situation was much more critical for the Small Probes because they had to be targeted precisely to complete their missions.

The Large Probe was launched in a direction so that it would enter near the equator on the day side of Venus. Separation was normal and the probe as an independent spacecraft silently pursued its path to Venus.

The internal timer counted the seconds to awakening its systems just before hurtling into the planet's atmosphere.

With the Large Probe separated successfully, the Bus was prepared for separating the three Small Probes. During the four days before scheduled release of the Small Probes, the Doppler residual uncertainty problem was confirmed as the effects of solar radiation. The change in aspect of the spacecraft during the prerelease maneuver changed the effects of solar radiation from those calculated earlier.

Twenty-two days before entry the Small Probes were successfully checked out by radio command. Two days later the Bus was reoriented to target the Small Probes to their entry points as shown in Figure 4.16–one on the day side at mid-southern latitudes (the Day Probe), the second on the night side, also at mid-southern latitudes (the Night Probe), and the third on the night side at high northern latitudes (the North Probe).

With the Multiprobe spacecraft oriented correctly and spinning at close to 48 rpm, clamps opened and the three Small Probes were released. The spin of the spacecraft and the precise timing of release directed the probes onto the trajectories required to reach the desired locations. In each probe its timer began counting the seconds to atmospheric entry.

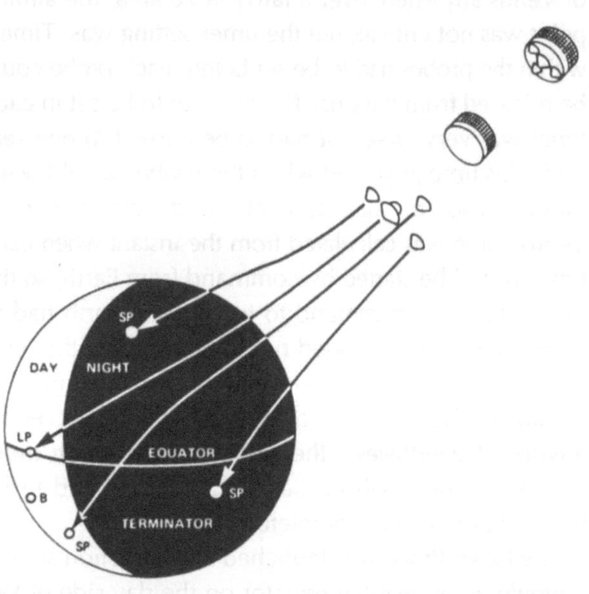

Figure 4.16. Target points for the probes; two in the daylit region and two at night and over a range of latitudes. (NASA/ARC)

After all probes had left, the Bus was maneuvered for its own atmospheric entry and was slowed slightly so that it would reach Venus a short while after the probes. The Bus did not have a heat shield and was expected to burn up within a few minutes. Two scientific instruments—ion and neutral mass spectrometers—would gather data about atmospheric composition between the top of the atmosphere and the 115 km (69 mi) level at which the Bus was expected to be destroyed.

One of the great challenges was to direct the Bus for entry into the atmosphere at as shallow a flight path angle as practical to extend its lifetime during its data gathering operation. But the trajectory should not be so shallow that the Bus would skip back into space off the top of the atmosphere without getting any atmospheric data. Ideally the Bus would penetrate to the 115 km level and then skip out again thereby gathering data along incoming and outgoing paths. But it was not possible to navigate so accurately. The risk would be too great that the depth of penetration needed would be missed.

The selected entry path was 9 degrees below horizontal at 200 km (125 mi) above the surface. The spin axis of the Bus had to be set with the angle of attack at 5 degrees so that atmospheric molecules would enter the scientific instruments. With all the probes and the Bus on their way to their targets the Orbiter approached its rendezvous with Venus.

The next big operational challenge occurred on December 4 when the Orbiter had to be injected into an elliptical path around Venus (figure 4.17) before the arrival of the probes. The maneuver had to take place when the spacecraft was hidden from Earth behind Venus and out of communication. Its solid-propellant rocket motor would slow the spacecraft sufficiently to be captured into an orbit around the planet. This was the first time such a rocket had been fired after being in space for seven months.

On December 2 the Orbiter started to maneuver for its insertion. The Orbiter had encountered some problems with the command memories on its way to Venus which could have led to serious difficulties in obtain-

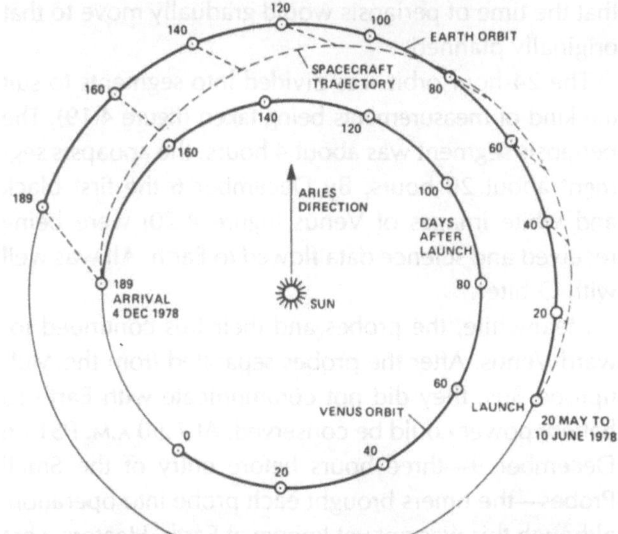

Figure 4.17. On arrival of the Orbiter at Venus, just before the arrival of the Multiprobe, the spacecraft had to be inserted into an elliptical orbit around the planet. (NASA/ARC)

ing a correct injection into orbit. High-energy cosmic rays had caused errors in the spacecraft's memories—changing ones to zeros and vice versa. These errors had occurred about once every two weeks or so, but fortunately at times when the command could be corrected or after the command had been executed. Had one occurred in the command timing sequence to fire the rocket motor for orbit insertion, it could have caused failure of the orbital insertion maneuver.

Countdown to ignition could alternatively be commanded through a series of commands for a sequence of small time delays whose sum would be the total time. Greater reliability would thereby be obtained from such a series of time delays because if any one of the commands for these time delays was changed by an anomaly, the overall effect on the ignition time would be minimal. Accordingly, on December 3 at 11:00 P.M. PST, the two command memories of the Orbiter were loaded with the sequence of commands needed to fire the orbit insertion motor at 7:58 A.M. on the following day.

At 7:51 A.M. on December 4 the Orbiter passed behind Venus and communications with Earth ended. Seven minutes later the orbit insertion commands stored in the spacecraft's memory fired the rocket motor which burned for almost 30 seconds to change the spacecraft's velocity by about 3780 kph (2349 mph). When the spacecraft emerged from behind Venus with its new velocity, three minutes had to elapse for the radio signals to travel the 56 million km (35 million mi) to Earth. Everyone waited for a receiver of the Deep Space Network to acquire the signal at the new frequency. When it was clear that the receiver had locked onto the spacecraft there was a big cheer; everyone knew that orbit had been achieved (figure 4.18).

At 8:30 A.M. the Orbiter's spin rate was adjusted to

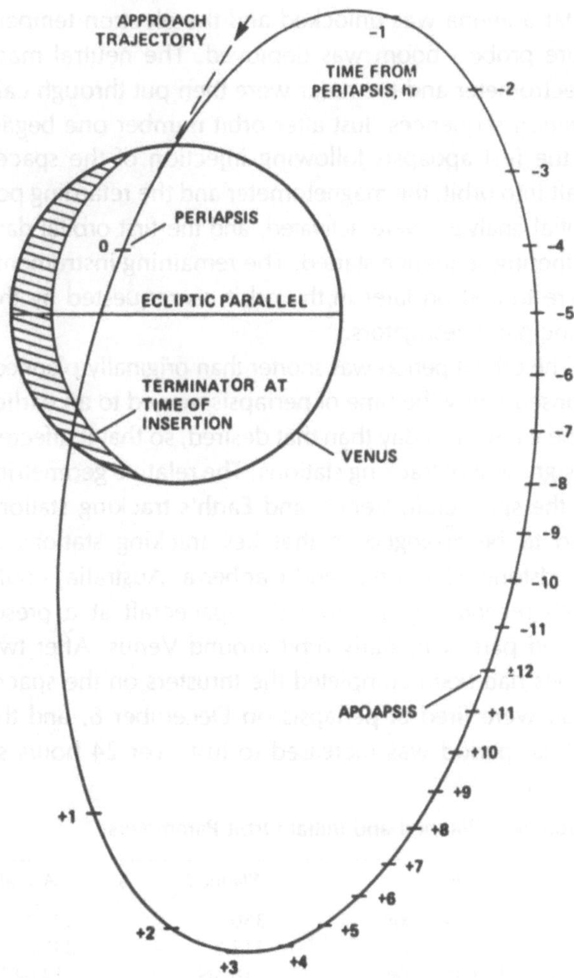

Figure 4.18. The orbit around Venus would have a period of 24 hours and be inclined considerably to the equatorial plane of the planet, as shown. (NASA/ARC)

15 rpm and the high-gain antenna was despun. Within the next few hours tracking data revealed the size of the orbit. Highly elliptical, the orbit was inclined 75 degrees to the equator of Venus (see table 4.2).

The Orbiter was maneuvered on December 5 to lower the periapsis from 378 km (234 mi) to 250 km (155 mi). Initial operations then followed a preplanned sequence. The spin rate was reduced and the spin axis adjusted to point toward the celestial poles. Then the high gain antenna was pointed to Earth and communication switched to it from the omni antenna. The infrared radiometer, the neutral mass spectrometer, and the electron temperature probe were turned on. The radar antenna was unlocked and the electron temperature probe's boom was deployed. The neutral mass spectrometer and the radar were then put through calibration sequences. Just after orbit number one began at the first apoapsis following injection of the spacecraft into orbit, the magnetometer and the retarding potential analyzer were activated, and the first orbital data gathering sequence started. The remaining instruments were turned on later in the orbit, as requested by the principal investigators.

The orbital period was shorter than originally planned. Consequently the time of periapsis moved to an earlier time each Earth day than that desired, so that it affected assignment of tracking stations. The relative geometries of the spacecraft, Venus, and Earth's tracking stations had to be arranged so that key tracking stations at Goldstone, California, and Canberra, Australia, could both receive signals from the spacecraft at a preselected part of its daily orbit around Venus. After two orbits had been completed the thrusters on the spacecraft were fired at periapsis on December 6, and the orbital period was increased to just over 24 hours so that the time of periapsis would gradually move to that originally planned.

Table 4.2. Planned and Initial Orbit Parameters

Parameter		*Planned*	*Actual*
Periapsis altitude,	km	350	378.7
	mi	217.5	235.3
Periapsis latitude,	deg	18.5N	18.647N
Periapsis longitude,	deg	203–223	207.990
Inclination,	deg	105	105.021
Period,	hr:min:sec	24:0:0	23 11.26

The 24-hour orbit was divided into segments to suit the kind of measurements being taken (figure 4.19). The periapsis segment was about 4 hours, the apoapsis segment about 20 hours. By December 6 the first black and white images of Venus (figure 4.20) were being received and science data flowed to Earth. All was well with Orbiter.

Meanwhile, the probes and their bus continued toward Venus. After the probes separated from the Multiprobe Bus, they did not communicate with Earth so battery power could be conserved. At 7:50 A.M. PST on December 9—three hours before entry of the Small Probes—the timers brought each probe into operation, although this was not yet known at Earth. Heaters were activated to warm the battery and the oscillators of the radio transmitters. Then the command unit started warmup and calibration for the three instruments on each of the probes. At 8:15 A.M. the command timer on the Large Probe initiated warmup of its battery and radio receiver for an uplink carrier frequency from Earth to provide the reference frequency for the downlink signal to Earth.

At 10:23 A.M., 22 minutes before entering Venus's atmosphere, the Large Probe began to transmit radio signals to Earth for Doppler tracking. The 22-minute interval was planned for the probes as a compromise between consuming precious battery power and providing the Deep Space Network stations with sufficient

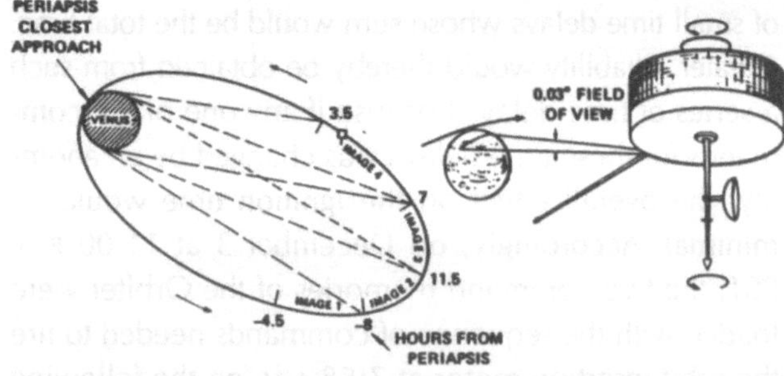

Figure 4.19. A cloud photopolarimeter carried by the Orbiter used the spacecraft's flight path around Venus coupled with rotation of the spacecraft to build up a picture of the planet in a series of scans. The instrument could make five images of Venus on each orbit around the planet. (NASA/ARC)

Figure 4.20. Soon after entering orbit, the Orbiter was returning excellent pictures of the cloud-shrouded planet showing the details of the ultraviolet markings. (NASA/ARC)

time to lock onto the signals before the probes started sending entry data. Within the next 11 minutes all the Small Probes came alive; first the North Probe, then the Day Probe, and finally the Night Probe. Seventeen minutes before it hurtled into the Venus atmosphere at 42,000 kph (26,000 mph), each probe started sending data.

The first radio signal to be received at Earth came from the Large Probe. It arrived at 10:27 A.M. PST on December 9. Three or four minutes later signals from the other probes reestablished communications with the Pioneer Mission Operations Center at NASA Ames Research Center, and data were arriving at Earth. By 10:45 A.M. all instruments on all the probes were operating.

The probes entered the appreciable atmosphere some 200 km (125 mi) above the surface of Venus, and an expected entry communications blackout occurred. It cut off the radio signals for about 10 seconds as each probe's meteoric plunge generated a radio-impervious plasma screen. The probes traveled more slowly after this blackout so the tracking stations had to acquire them again at a slightly different Doppler-shifted radio frequency. The Deep Space Network successfully found the signals again for all the probes.

Now the most exciting part of the mission began. Enormous pressure and intense heat coupled with chemical corrosion were the great environmental challenges in designing and building the probes. As the probes plunged deep into the alien atmosphere, engineers anxiously awaited results over the telemetry. Although the probes had withstood many tests on Earth, there was the possibility that the environment of Venus would spring some surprises. Table 4.3 summarizes the time of some important events occurring during the entry of all the probes.

The Large Probe slowed in 38 seconds from 41,800 to 727 kph (26,000 to 452 mph). Data were temporary

Table 4.3. Important Entry Events for the Pioneer Probes

	Time at Spacecraft (hr.min.sec PST)			
Event	*Large*	*North*	*Day*	*Night*
End of coast timing	10:24:26	10:27:57	10:30:27	10:34:08
Initiate telemetry	10:29:27	10:32:55	10:35:27	10:39:08
200-km (124-mi) entry	10:45:32	10:49:40	10:52:18	10:56:13
Radio blackout began	10:45:53	10:49:58	10:52:40	10:56:27
Signal locked on	10:46:55	10:50:55	10:53:46	10:57:48
Jettison parachute	11:03:28	(none on Small Probes)		
Impact with surface	11:39:53	11:42:40	11:47:59	11:52:05
Signal ended	11:39:53	11:42:40	12:55:34	19:52:07

Bus entry (200 km; 124 mi)	12:21:52
Bus signal ended (110 km; 68 mi)	12:22:55

Durations (minutes)	*Large*	*North*	*Day*	*Night*
Descent time (entry to impact)	54:21	53:00	55:41	55:52
Blackout time (signal loss to relock)	62	57	66	81
Time on parachute (Large Probe only)	~17.07			
Surface operations (impact to signal end)	none	none	67:37	02

NOTE: Earth received times were approximately 3 minutes later than the above spacecraft times.

stored in its onboard memory for transmission after the radio blackout. Its parachute opened at 10:45 A.M. to further slow its speed. Its forward aeroshell heat shield was jettisoned to expose all apertures and windows for the probe to begin its descent phase of operations. Forty-three seconds after entry, at an altitude of about 66 km (40 mi), all instruments were sending data to Earth. Seventeen minutes later, at 10:03 A.M. and an altitude of 45 km (28 mi) above the surface, the parachute was jettisoned and the probe continued to descend, slowed by the dense atmosphere. It reached the surface of Venus in close to 36 minutes, hitting it at about 32 km/hr (20 mph) near the equator of Venus on the day side at 11:39 A.M. This was 54 minutes after entering the atmosphere. Radio transmission ended at impact.

Five minutes before the peak deceleration of the Small Probes, each command unit ordered the blackout format for storage of spacecraft data in an internal memory. As with the Large Probe these stored data were transmitted later during the descent.

The Small Probes entered the atmosphere within a few minutes of each other between 10:50 and 10:56 A.M. At 10:51 A.M. the window for the nephelometer was opened on the North Probe and it began gathering data on cloud layer locations and densities. The housing doors of the atmospheric structure and net flux radiometer opened and these instruments started sending data about the thermal structure of the atmosphere. Within the next six minutes similar sequences had started on the two other Small Probes.

All the Small Probes took about the same time as the Large Probe to reach the surface. As the probes penetrated deeper into the atmosphere the signals received at Earth from them weakened. At 16.4 minutes after entry and an altitude of about 30 km (18 mi), the rate of data transmission was reduced to avoid missing any data about the lower atmosphere. From that point on the three probes descended into increasingly dense atmosphere, impacting the surface at 36 km/hr (22 mph) 53–55 minutes after their entry times. Unlike the Large Probe, the small probes retained their heat shields to the surface.

The North Probe landed at 11:42 A.M. in darkness near northern polar regions. The Day Probe went into the southern hemisphere on the day side and landed at 11:48 A.M. The Night Probe went down in darkness to reach the surface in the southern hemisphere at 11:52 A.M. (figure 4.21). While signals from the North Probe

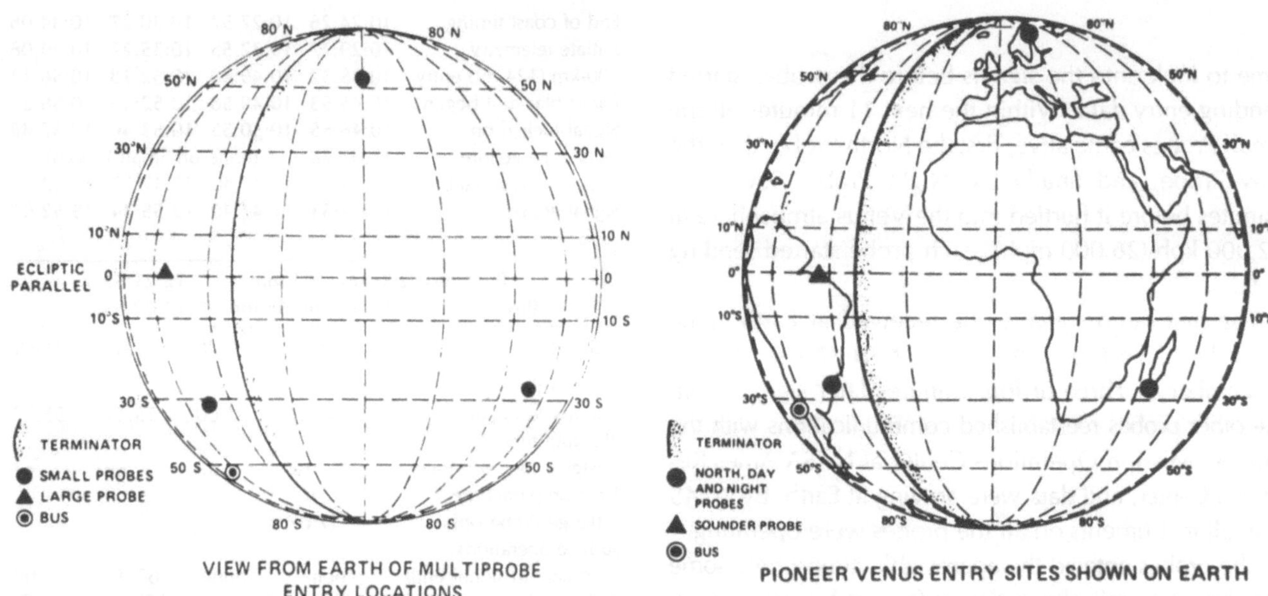

Figure 4.21. The entry sites of the Pioneer Venus probes on Venus are compared with similar locations on Earth. (NASA/ARC)

and the Night Probe ended at the time of impact, transmissions continued for 68 minutes from the Day Probe on the surface. Engineering data showed its internal temperature climbing steadily to reach 126° C (260° F) when its batteries became depleted and its transmission ended. The internal pressure monitors showed that the internal pressure rose as would be expected as a result of increasing temperature. There was no sign of leakage through the seals after the impact. Table 4.4 shows the locations on Venus where the probes impacted and the conditions at the impact points. All were close to the targeted locations.

On December 8 the Bus had been oriented to its final entry angle; its instruments were calibrated and the cap covering the inlet to the neutral mass spectrometer was released. Entry took place at 12:22 P.M. on December 9, about 96 minutes after entry of the first probe, and 88 minutes after all the probes had entered. The Bus entered on the day side at a high southern latitude. Table 4.5 gives the entry point of the Bus at 200 km (125 mi) altitude, and the locaions of where the Sun and the Earth would appear directly overhead to an observer on Venus.

The Bus had no heat shield and was expected to burn up within two minutes. It transmitted data on the composition of the very high atmosphere, including the region where the ionosphere is most dense, a region which could not be explored by the other probes. The Bus entry was delayed from the entry of the probes so that it could provide a radio reference for the long baseline interferometry experiment.

When the Bus burned up at 12:23 P.M. the exciting entry phase of the mission was concluded. In only 1 hour and 38 minutes, the probes and the Bus had gathered a mass of new information about the atmosphere on Earth's errant twin. During the following few days

Table 4.4. Pioneer Venus Multiprobe Impacts on Venus

Probe Name	*Latitude (deg)*	*Longitude (deg)*	*Solar Zenith Angle (deg)*	*Local Venus Time*
Large	4.4N	304.0	65.7	7.38
North	59.3N	4.8	108.0	3:35
Day	31.3S	317.0	79.9	6:46
Night	28.7S	56.7	150.7	0:07

Table 4.5. Pioneer Venus Bus Entry and Location of Sun and Earth Subpoints

	Latitude (deg)	*Longitude (deg)*	*Solar Zenith Angle (deg)*	*Local Venus Time*
Bus entry at 200 km	37.9S	290.9	60 7	8:30
Subsolar	0.5S	238.5	0	12:00
Subearth	1.6S	1.7	123.1	3:47

scientists completed preliminary initial analyses of the data and announced some startling and unexpected discoveries. Then the mission settled down to the equally fascinating but lengthy process of observing Venus from the Orbiter over a period of several Venus sidereal days.

The probes determined the composition and abundances of major, minor, and noble gas species in the lower mixed atmosphere, and in the uooer, diffusively-separated atmosphere. They discove d that the isotopic ratios of certain selected speci of gases were quite different from those in Earth's a nosphere. Their instruments measured temperature, pr sure, and density globally and vertically from the su ce through the clouds into the upper atmosphere. Sou es and sinks of solar and infrared radiation were loc d in the lower atmosphere and clouds under daytim nighttime, low latitude, and high latitude conditior The runaway greenhouse effect was confirmed as b g a partial explanation of the high surface temper ure. The probe data showed the structure of the clou , globally and vertically—their layers, particle size d ribution, composition, and optical properties. Lov ltitude glows, possibly lightning, were observed ir /enus's atmosphere.

The probes obtained measurements show how the wind velocity varies with altitude at fo locations, and measured global winds at the cloud to Observations were obtained of composition, state pr erties, clouds, thermal balance, and winds, for integr on into a general meteorological model that could b ompared with meteorologies of other planets. These esults are discussed further in later chapters.

The probes and their instruments stood the rigors of the descent into Venus's atm here remarkably well. One possible source of troub had been anticipated. As discussed earlier, to prevent droplets condensing from the clouds onto the inlet to the mass spectrometer, a heater coil had been placed around the inlet. However, despite this precautionary measure, the inlet did become blocked and this blockage was manifested as a change in the amount of gas entering the instrument. Later in the descent, when the temperature had risen, the amount of sulfur measured peaked, presumably as a large drop of sulfuric acid boiled off.

Fortunately, the blockage enhanced rather than detracted from the science results. A sample of gas had been taken earlier by the mass spectrometer and carbon dioxide was then stripped from it. The sample was next introduced into the ion source to measure isotopic ratios. This ratio experiment happened to coincide with the blockage of the gas inlet port which prevented any other gas from contaminating the isotopic ratio sample being analyzed at that time.

There were some mysterious events during the descent of all the probes and these were apparent in engineering data and science data gathered at about the same altitude. The first strange signs came from the temperature sensors of an atmospheric structure experiment. Soon afterward net flux radiometers' external sensors on the North, Day, and Night Probes suddenly failed, all at about the same altitude. The data from other scientific instruments and from engineering transducers contained unexpected values that were measured just before, during, and after these failures. These anomalies are summarized in table 4.6.

The net flux radiometer of each of the three Small Probes, and the temperature sensor on all four probes stopped working between 12 and 14 km (7.5 and 9 mi) altitude, at a time when other spacecraft functions also behaved mysteriously. Moreover, the effect was observed at all locations on the planet at the same range of altitudes. At about 14 km (8.7 mi) the temperature sensors of every probe produced unusual data, as did thermocouples on the heat shields of the Small Probes and a thermistor which measured transition joint temperature. And at this same time the net flux radiometer stopped sending data.

There was and to this day is no simple explanation why all these different instruments failed at precisely the same time. Minor manufacturing differences were

Table 4.6. Strange Occurrences Experienced by the Probes

	Probe			
Unusual Event	*Large*	*North*	*Day*	*Night*
Temperature sensor data interrupted	✓	✓	✓	✓
Changes and spikes in pressure data	✓	✓	✓	✓
Abrupt changes in cloud particle size laser alignment monitor	✓			
Apparent failure of net flux radiometer fluxplate temperature sensors		✓	✓	✓
Abrupt changes and spikes in data from net flux radiometer		✓	✓	✓
Decrease in the intensity of the beam returned to the cloud particle size spectrometer	✓			
Steady increase in flux readings of the infrared radiometer	✓			
Change in the indicated deployment status of the atmosphere structure temperature sensor and net flux radiometer booms		✓	✓	✓
Erratic data from two thermocouples embedded in the heat shield		✓	✓	✓
Erratic data from a thermistor measuring junction temperature of the heat shield thermocouples		✓	✓	✓
Noise in the data from the infrared radiometer	✓			
Spikes in the data monitoring the ion pump current of the mass spectrometer analyzer	✓			
Abrupt decrease of current in the power bus	✓			
Slight variation of current and voltage levels in the power bus		✓	✓	✓
Slight offsets or jumps in the values for temperatures of the forward and aft instrument shelves and in the measurements of the internal pressure		✓	✓	✓
Jumps in the receiver (transponder) static phase error	✓			
Spikes in the receiver automatic gain control	✓			
Spurious reading from thermocouples that had been dropped from the Probe in its heat shield and should no longer be producing data	✓			

to be expected in similar instruments but it was most unlikely that all instruments would fail at the same level in the atmosphere. A cause other than equipment failure seemed more reasonable.

It was clear from the temperature data that the sensors did not physically break because when each probe landed the final readings indicated the correct surface temperature of about 500°K and the electrical resistance in the wires of the sensor was as expected. Earlier when each probe was in the clouds there was partial shorting, varying from probe to probe, of the insulation of the fine-wire sensors. However, this cleared as the probes descended, and was not a problem at lower altitudes.

Nevertheless, the probes performed remarkably well in the extremely inhospitable atmosphere of Venus. Much new science and engineering data were gathered, and the technology had been proved for penetrating planetary atmospheres and gathering data at high temperatures and pressures. This new technology opened the way to explore planetary atmospheres of the giant worlds of the outer Solar System.

Meanwhile, preliminary science discoveries came from the Orbiter in its long orbital mission. Global winds were measured at the cloud tops to supplement the wind velocity at many different altitudes measured by the probes. An overall picture of circulation in Venus's atmosphere began to emerge. The nature of the interaction of the solar wind with Venus was clarified, and information about the locations of the bow shock and the ionosphere was obtained. The rates at which particles enter the ionosphere and energy is absorbed in it were determined. Radar images of the surface, and altimetry measurements to reveal topography, were obtained. Many geologic features were identified, such as rifts, mountains, continental masses, and great plains. Local gravity perturbations were measured. When these were combined with the radar observations, Venus's surface and interior seemed more like those of Earth than those of Mars.

The Orbiter data confirmed that Venus has little, if any, intrinsic magnetic field, and the data established upper limits to the magnetic moment of the planet, much smaller than on previous missions. The Orbiter's radar mapping (figure 4.22) suggested that the topography of Venus might be similar to that of Earth, with high mountainous regions and extensive relatively flat areas. The scans by the radar mapper showed that in a region of Venus previously unexplored by radar from Earth

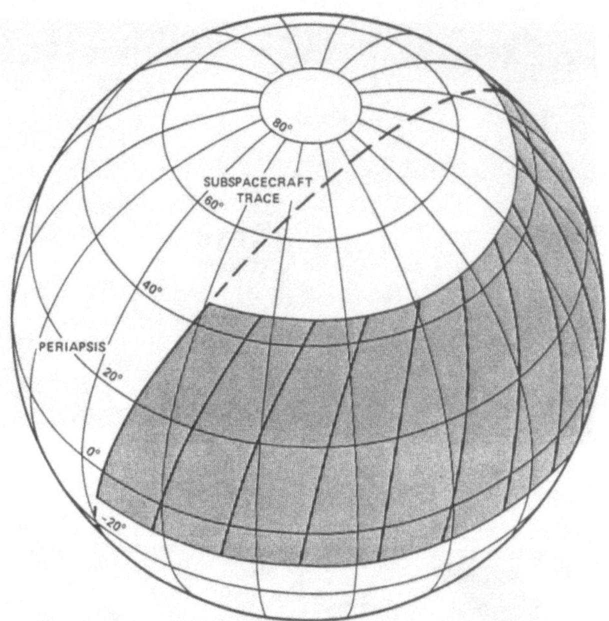

Figure 4.22. The radar mapper carried by the Pioneer Venus Orbiter produced radar maps of the surface from a number of scans, as shown in this diagram. (NASA/ARC)

much of the surface was relatively flat, similar to Earth's surface, and quite different from the rough, cratered surfaces of Mars, Mercury, and the Moon (figure 4.23).

A serious setback occurred to the mapping when the radar instrument stopped working. Teams of scientists and engineers analyzed the situation and tried several remedies, but to no avail. This failure was extremely disappointing because the images had started to show intriguing details of the surface. No answer was forthcoming and the instrument refused to gather data. Regretfully, scientists had to turn it off. But a month later, when the radar was turned on again, it worked. The problem appeared to be overheating caused by running the instrument for periods longer than 10 hours. Henceforth the instrument was operated for a short time only on each orbit and was turned off the rest of the time. Under these conditions the radar gradually returned to normal performance within about ten days, and it operated satisfactorily for the rest of the mission. Although a month of radar data was lost, the areas of Venus missed during the period the radar was inactive were covered later in an extended mission. Another disappointment came on orbit 70, when the infrared

Figure 4.23. The preliminary and first topographic map of Venus revealed two large, continent-sized, highland areas. The black section is a part of the planet not covered when the radar equipment failed for a period. This area was covered later during the extended mission of the Orbiter. (NASA/ARC)

radiometer failed and could not be brought back into operation. It is believed that the problem was in the power supply. It remains the only permanent instrument failure in the program.

The initial altitude of periapsis for the Orbiter had been chosen to be high enough for drag on the spacecraft to be negligible during the first orbit. At the beginning of the orbital mission there was not much information about the upper atmosphere. As more knowledge was built up about this region by trial-and-error it became practical to make corrections to the perapsis and lower it to the desired 150 km (93 mi) abovethe mean surface of Venus. Orbital parameters for the nominal mission are given in table 4.7.

The location of the periapsis relative to the planet is affected by perturbations from the Sun. This required using the thrusters to maintain the orbit within predetermined limits. Without corrections to the orbit, the Sun's gravity would raise the altitude of the periapsis. To keep the periapsis within the range of alitudes required, periodic corrections were made throughout the mission.

During the first few weeks in orbit the spacecraft's periapsis was lowered to 150 km (93 mi) before it passed from the dayside to the nightside of Venus. The lower atmospheric density on the nightside permitted the periapsis to be lowered several times to 142 km (88 mi) for sampling even deeper into the atmosphere. Because the local time of periapsis increased by 1.6° per day, the Orbiter's sampling at periapsis moved from the dayside of Venus, then across the evening terminator, and later into the night side. Later still the periapsis crossed the morning terminator and the spacecraft sampled the dayside again. Data at periapsis and along the orbit were thus obtained for all Venus local times in a period of 224.7 Earth days. Because of the retrograde rotation of the planet the longitude of periapsis moved at 1.48° per Earth day (i.e., per orbit) so that it took 243 Earth days to sample all longitudes on the surfaces of the planet.

Table 4.7. Orbital Parameters for the Orbiter's Nominal Mission

Parameter	*Value*
Periapsis km	150–2(0
mi	93–1:4
Apoapsis km	66,90[illegible]
mi	41,57[illegible]
Eccentricity	0.843
Average Period hr	:4.03
Inclination to Equator deg	1(5.6
Periapsis Latitude deg	7.0N
Periapsis Longitude deg (for Orbit 5)*	1'0.2

The nominal mission of the Orbiter was completed on August 4, 1979. Enough propellant remained to keep the spacecraft active for at least another two sidereal periods of Venus; i.e., for another 486 days. This extended mission provided a tremendous scientific bonus from a relatively inexpensive planetary mission.

Periodic control of the orbit continued until midsummer 1980, when the periapsis altitude was allowed to rise slowly, initially at a rate of 400 km (250 mi) per 243 days and at only 225 km (140 mi) per 243 days by 1984. The apoapsis fell at an identical rate so that the period of the orbit remained constant.

The extended mission had two operational phases. In the first, periapsis remained within Venus's atmosphere so that the spacecraft could continue to gather atmospheric data. During the second phase large regions of the dayside bow shock and the nightside ionosphere that could not be investigated in the nominal mission and the first phase of the extended mission became accessible about one month twice each Venus year. This phase also provided an opportunity to track the spacecraft for improved estimates of the lower-order gravity field of Venus, since atmospheric drag was virtually negligible.

5

THE VEILS OF VENUS

While in recent years there have been tremendous improvements in techniques for Earth-based observations of Venus, the major breakthroughs in understanding Earth's errant twin, have developed from the missions of spacecraft to the veiled planet.

At press time, there have been 22 of these missions to Venus, 17 by the Soviet Union and 5 by the United States of America. Spacecraft that were unsuccessful in leaving the vicinity of Earth are not included. The missions are summarized in the Appendix.

The veils of Venus are the various regions of the planet's environment between its surface and space. Outermost is the magnetosphere, which holds off the solar wind and prevents it from plunging to the solid surface of the planet as happens on Earth's Moon. Into this magnetosphere stretches the exosphere, where individual molecules and atoms can travel for many miles without encountering each other. There, atoms and molecules follow ballistic trajectories. In the lower part of this region the atmosphere becomes dense enough for collisional processes to become important, and in this region incoming solar radiation intercepts with sufficient numbers of atoms and molecules in the atmosphere to produce an ionosphere of charged particles.

Closer to the planet are regions where solar radiation at longer wavelengths heats the atmosphere and creates a thermosphere. Still nearer to the surface, the atmospheric density is such that weather conditions arise in a troposphere, where clouds form.

The missions to Venus have provided much-needed information about these regions, which has enabled scientists to compare them with equivalent regions or comparable regions on other planets and to become aware of striking differences between Venus and Earth.

Of particular importance has been the discovery that Venus, although it lacks a significant intrinsic magnetic field, is able to hold off the solar wind by a field generated as a result of the solar wind streaming past the planet. Also, the atmospheric probes to Venus have revealed the structure and the composition of the cloud layers and identified their location in the atmosphere. Circulation patterns of the massive atmosphere have been established and wind velocities plotted from high altitudes down to the surface. Now meterologists have much information about an atmosphere 100 times as dense as Earth's on a planet that rotates extremely slowly compared with Earth. This provides useful comparisons with the Martian atmosphere (which is one hundred times less dense than Earth, but on a planet rotating at the same rate as Earth), and with the Jovian atmosphere (where there is rapid rotation but no solid surface to affect an enormously deep atmosphere).

The expeditions to Venus have thrown new light on conditions within three major regions of Venus's environment: the lower atmosphere, the upper atmosphere and its ionosphere, and the magnetosphere. The missions of spacecraft to Venus confirmed that the appreciable atmosphere begins at an altitude of 250 km (155 mi), where it has a density of 10^{-15} gm/cc. This density increases to 10^{-10} gm/cc at 125 km (78 mi). The turbopause is a region where atmospheric mixing replaces atmospheric layering by molecular weight. On Venus it begins at an altitude of 144 km (90 mi); on Earth it is at 100 km (62 mi). Above the turbopause of Venus Pioneer's instruments detected hydrogen, oxygen, carbon dioxide, argon, helium, nitrogen, and carbon monoxide.

The exosphere, from which gas molecules escape into space, starts on Venus at 160 km (100 mi), just a little above the turbopause. The base of Earth's exosphere, by contrast, is at 550 km (341 mi), far above Earth's turbopause. Temperature in Venus's atmosphere at 250 km (155 mi) was measured as 27° C (80° F), and at 100 km (62 mi) it was −93° C (−136° F). Near the surface the temperature reached 460° C (860° F).

The region of maximum ion density in Earth's ionosphere is at an altitude of 300 km (186 mi), far above Earth's turbopause. On Venus the maximum ion density was recorded at 145 km (90 mi), close to the turbopause. Ion density decreased very rapidly below that altitude. At the maximum region the most abundant ions in Venus's atmosphere were molecular oxygen and carbon dioxide, but atomic oxygen became abundant above the ion maximum region, increasing from there to the top of the ionosphere. The planet has a nighttime ionosphere despite a night that is 58 times longer than Earth's.

About 75 percent of the sunlight reaching Venus is actually reflected back to space by the clouds, so that all things being equal, Venus might be expected to be cooler than Earth even though closer to the sun. The planet's dense lower atmosphere and searing surface heat seem to be due in part to a runaway greenhouse effect (figure 5.1). In the greenhouse effect, solar radiation enters the atmosphere relatively easily but is reradiated to space with great difficulty, so that the trapped radiation increases the temperature of the planet's surface and its atmosphere. Of the solar radiation that penetrates the atmosphere, about 15 percent is absorbed in the clouds, 3.75 percent in the atmosphere above the clouds, 3.75 percent in the lower atmosphere, and 2.5 percent at the surface. By contrast, aproximately 30 percent of the solar radiation received by Earth reaches the surface. The greenhouse effect has

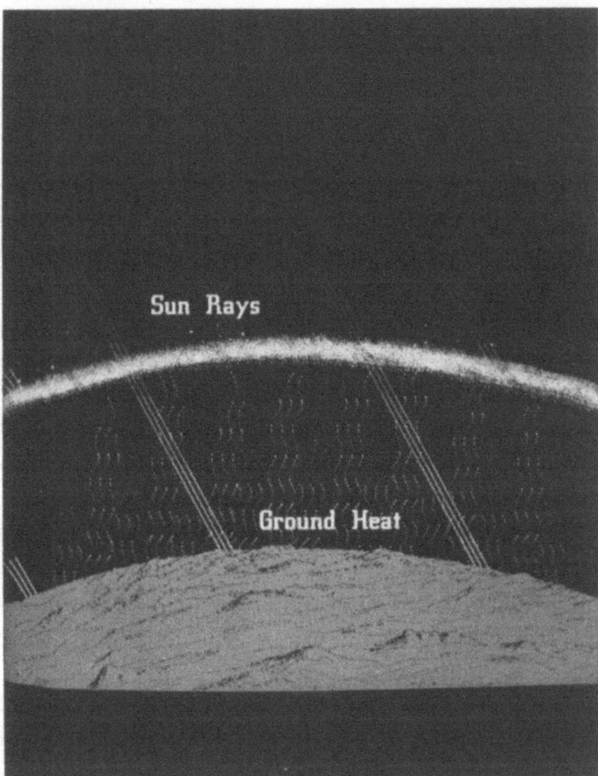

Figure 5.1. The runaway greenhouse effect of Venus's atmosphere traps incoming energy from the Sun and prevents its reradiation into space. As a result the atmosphere and surface of Venus have become extremely hot by terrestrial standards.

thus pushed Venus to much higher temperatures than Earth.

Two materials identified by spacecraft as present in Venus's atmosphere are mainly responsible for holding the heat in the atmosphere. Carbon dioxide traps much of the heat but not enough to maintain the high temperatures on Venus. Pioneer Venus discovered that the atmosphere of Venus contains water vapor and possibly sulfur particles to assist the carbon dioxide in trapping heat and leading to the runaway greenhouse.

Atmospheric pressures of between 90.5 and 91.5 bars were measured at the surface. (One bar is approximately the pressure of Earth's atmosphere at sea level.) Composition of the atmosphere as measured by Pioneer Venus is about 97 percent carbon dioxide, 1–3 percent nitrogen, 250 ppm (parts per million) helium, 6–250 ppm neon, and 20–200 ppm argon. Other constituents detected below the cloud layers were water vapor, 0.1 to 0.4 percent, sulfur dioxide 240 ppm, and oxygen 60 ppm. There was indirect evidence that sulfuric acid and elemental sulfur particles are also present in the clouds.

A haze layer, about 10 km (6 mi) above the clouds, covered the whole of the planet and was thicker over the polar regions. An opaqueness to longer infrared wavelengths over the poles suggested the presence of water vapor or ice crystals, like those in cirrus clouds.

The ultraviolet patterns observed in the clouds of Venus by Mariner 10, and faintly on Earth-based photographs, posed some important questions about circulation in the atmosphere. The four-day period of rotation of these markings appeared to be well established from eight days of pictures obtained by Mariner 10. The question, however, was whether this rotation represented actual movement of masses of air or the manifestations of wave motions in the atmosphere. Pioneer Venus Orbiter significantly extended observations of the ultraviolet patterns in the clouds of Venus over a period of hundreds of days. This enormous improvement in the number of observations enabled scientists to conclude that the air is moving in bulk at a speed of about 100 meters per second at the cloud tops, which is about 60 times faster than the planet's surface rotates. An atmosphere rotating faster than its planet rotates, even in equatorial regions, is known as a superrotating atmosphere. Earth, by contrast, has a subrotating atmosphere except at high latitudes. The high angular momentum of the Venusian atmosphere is believed to derive from the transfer of angular momentum from the planet itself or from the Sun.

The various probes into the atmosphere showed that this high speed continues through the cloud layers, but below the clouds the speed falls off rapidly and then continues to decrease more gradually until it becomes very small at the surface. However, there are some cyclic wave motions, and it appears that large features, especially the Y and C markings, may be a type of waves moving around the planet, because markings move at a speed which is different from the speed of the winds. If a Y marking (figure 5.2) traveled at local wind speeds the marking would disintegrate over a couple of rotations, because at times high-speed mid-latitude jets are present. All four U.S. probes and some Soviet probes measured the same rapid westward motion of the clouds, with little or no motion to or from the polar regions.

The atmosphere has now been investigated by many instruments carried by orbiters, and has been sampled by probes and landers. Atmospheric physicists classify regions of a planet's atmosphere by the temperature

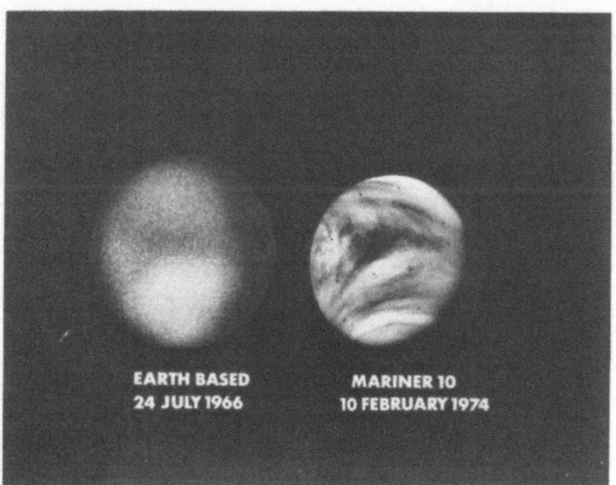

Figure 5.2. The ultraviolet markings of Venus were faintly discerned in Earth-based photographs but shown in much more detail by the first images obtained from spacecraft. (NASA/JPL)

and how it changes within a region. In the troposphere, for example, temperature decreases with increasing altitude. In the thermosphere, where incoming short-wavelength solar ultraviolet radiation is absorbed into the atmosphere, temperature increases with increasing altitude. The stratosphere is a region where temperature remains fairly constant, but increases slightly with increasing altitude and the atmosphere is fairly stable and layered. Earth has a well-defined stratosphere, Venus does not. (See figure 5.3).

The lowest region of the atmosphere of Venus is the troposphere, which extends to about 45 km (28 mi) above the surface. The temperature decreases vertically from 480° C (895° F) close to the surface to −23° C (−10° F) at the top of the cloud layers. In this region the thermal inertia of the atmosphere is so great that the temperature does not vary diurnally, even without winds to transport heat between various zones. The major circulation pattern may be a large, ponderous

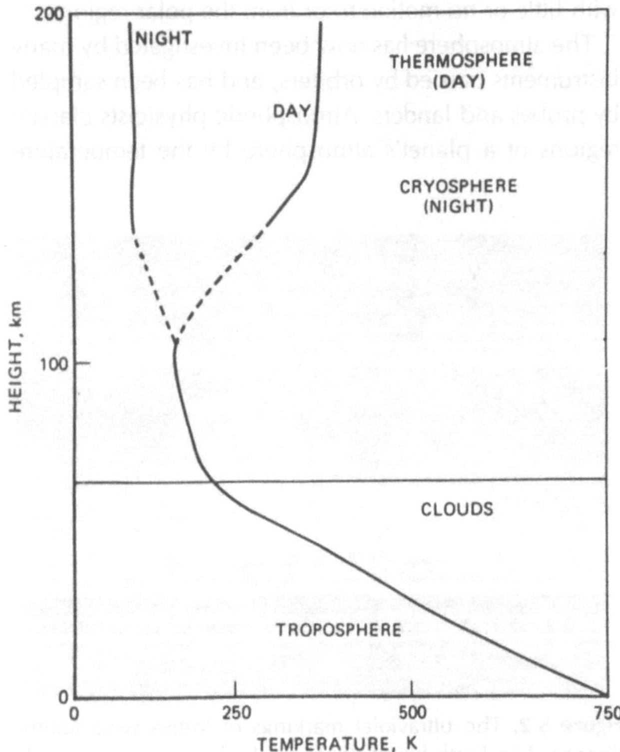

Figure 5.3. Regions of the atmosphere of Venus are identified in this diagram. (NASA/ARC)

Hadley cell with the atmosphere rising at the equator, moving slowly toward the poles, and descending again at high latitudes to move back slowly toward the equator. This type of thermally driven atmospheric circulation was first proposed for Earth's atmosphere by George Hadley in 1735 as an explanation for the Trade Winds. Heated atmosphere rises near the equator, flows aloft toward the poles, descends at high latitudes and flows back toward the equator near the surface. How this slow meridional flow on Venus changes to the high-speed zonal flow at the cloud tops is not yet fully understood.

The many differences between the atmospheres of Venus and Earth make it very difficult to understand what is taking place in the atmosphere of Venus on the basis of what we have discovered about our own. Perhaps circulation in the very dense atmosphere of Venus is more like the circulation in Earth's oceans than that in Earth's atmosphere. The pressure at the mean surface of Venus is, in fact, equal to the pressure 1 km (0.6 mi) beneath the surface of Earth's oceans.

An interesting fact about Venus is that because of its proximity to the Sun it receives nearly twice as much solar flux as does Earth but, as already mentioned, the atmosphere of Venus absorbs only about 55 percent as much solar energy as the Earth. This is because the albedo of Venus is much greater; its clouds reflect back into space about 75 percent of the incoming radiation. Earth, by contrast, reflects only 30 percent of the incident solar radiation. While the solar radiation received by Venus does not readily reach the surface, solar radiation penetrates to Earth's surface with relative ease. Heating due to absorption of solar radiation at the surface of Earth becomes the driving force for circulating the terrestrial atmosphere. On Venus, however, solar radiation is absorbed at high levels, and the driving forces for atmospheric circulation originate in the higher atmosphere of the cloud layers and above them. In addition, Venus has no seasonal effects on atmospheric circulation because its axis is only slightly tilted and its orbit is very nearly circular.

The major regions of the atmospheres of Earth and Venus are compared in figure 5.4. On the day side of Venus, the temperature of the upper atmosphere is not nearly so hot as Earth's upper atmosphere, where tem-

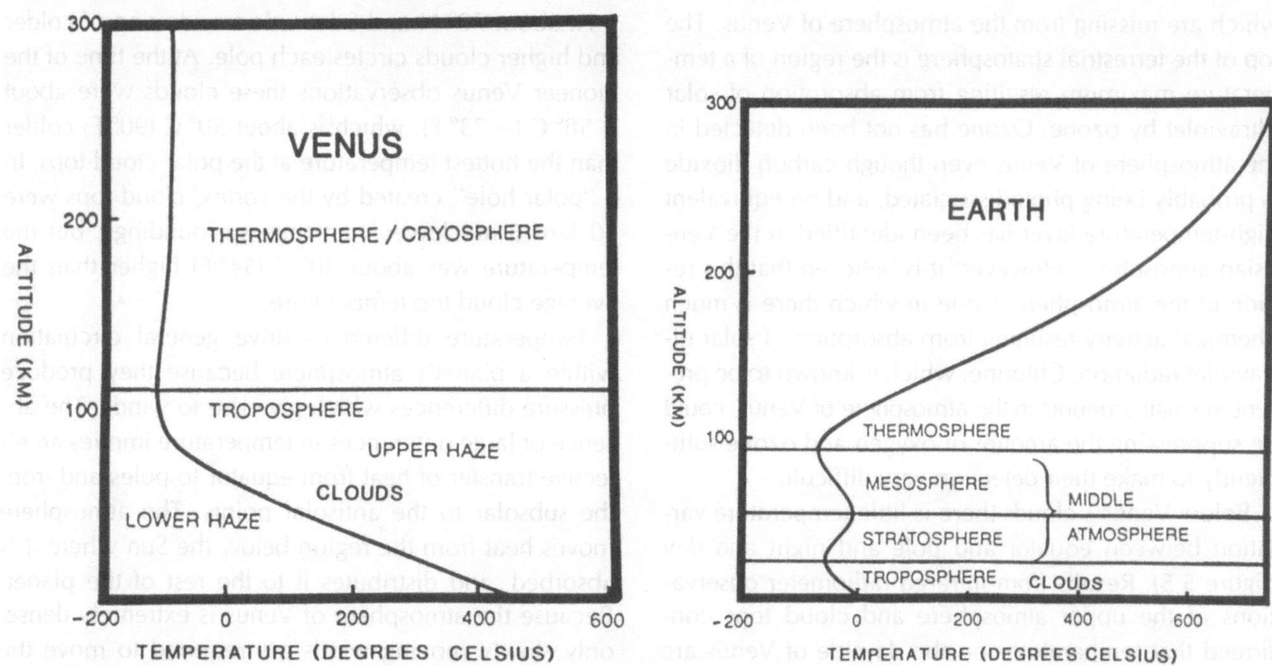

Figure 5.4. Atmospheric regions of Venus and Earth are compared on these diagrams.

peratures are 430° to 730° C (800 to 1316° F) at sunspot minimum. The heating of the upper atmosphere results from the processes forming the ionosphere by absorption of solar ultraviolet radiation at very short wavelengths. The upper atmosphere of Venus maintains a cooler temperature than Earth's upper atmosphere, even with twice the flux of incoming solar radiation. An exciting discovery was the enormous change in temperature between day and night in the upper atmosphere of Venus, with very low temperatures on the nightside. This region of the upper atmosphere of Earth is called the thermosphere (hot sphere). But this term cannot be used for the similar region of the night and day hemispheres of Venus's atmosphere. The cold region of the upper atmosphere of Venus at night has been named the cryosphere, or cold sphere.

Although the Sun does not heat the night atmosphere, some heat flows to it from the sunlit hemisphere. Also, on the nightside, heat must be flowing upward into the cryosphere from the warmer lower region of the troposphere in which the temperature does not vary from day to night. The high-altitude temperature gradient between day and night occupies little more than 20 to 30 degrees of longitude—essentially the zones of twilight on the planet. There is still no generally accepted theory to account for this anomalous temperature structure of Venus's atmosphere.

The nightside upper atmosphere is so cold that atmospheric pressure decreases rapidly with increasing height, so that, at equivalent altitudes, pressure in the night atmosphere is much less than that on the sunlit hemisphere. This pressure difference causes strong winds to blow from day to night, the presence of which has been confirmed indirectly.

The tropopause is the upper boundary of the troposphere. On Earth the boundary is about 15 km (9 mi) above sea level. On Venus it is at 110 km (68 mi) above the mean surface; namely, about the level of the cloud tops. Pressures at the tropopauses of Earth and Venus are similar, but the heights are quite different because of the different masses of the atmospheres and their surface pressures. In Earth's atmosphere the region above the tropopause, known as the middle atmosphere, consists of the stratosphere and the mesosphere, both of which are missing from the atmosphere of Venus. The top of the terrestrial stratosphere is the region of a temperature maximum resulting from absorption of solar ultraviolet by ozone. Ozone has not been detected in the atmosphere of Venus even though carbon dioxide is probably being photodissociated, and no equivalent high-temperature layer has been identified in the Venusian atmosphere. However, it is believed that this region of the atmosphere is one in which there is much chemical activity resulting from absorption of solar ultraviolet radiation. Chlorine, which is known to be present in small amounts in the atmosphere of Venus, could be suppressing the amount of oxygen and ozone sufficiently to make their detection very difficult.

Below Venus's clouds there is little temperature variation between equator and pole and night and day (figure 5.5). Results from infrared radiometer observations of the upper atmosphere and cloud tops confirmed that temperatures on the dayside of Venus are very nearly the same as those on the nightside. Other radiometer measurements at the poles appeared to confirm theories of a downward-moving polar vortex. But the belt of atmosphere above the cloud tops at the poles proved to be about 10° C (18° F) hotter than similar regions at the equator. This contradicted results from earlier missions which showed the poles as 10° C cooler than the equator.

Figure 5.5. The temperature on Venus does not vary significantly between day and night or from equator to poles.

At about 70° N and S latitude a wide ring of colder and higher clouds circles each pole. At the time of the Pioneer Venus observations these clouds were about −58° C (−73° F), which is about 50° C (90° F) colder than the hottest temperature at the polar cloud tops. In a "polar hole" created by the vortex, cloud tops were 10 km (6 mi) lower than their surroundings, but the temperature was about 30° C (54° F) higher than the average cloud top temperature.

Temperature differences drive general circulation within a planet's atmosphere because they produce pressure differences which give rise to winds. The absence of large differences in temperature implies an effective transfer of heat from equator to poles and from the subsolar to the antisolar points. The atmosphere moves heat from the region below the Sun where it is absorbed, and distributes it to the rest of the planet. Because the atmosphere of Venus is extremely dense, only slowly moving winds are required to move the heat. This dense atmosphere also reduces the rate at which temperature rises or falls as the input of solar heat varies.

Scientists were surprised to find that the lower atmosphere of Venus consists of very stable layers with hardly any mixing between them. This region is more like Earth's stratosphere than troposphere. There was no convective activity observed in a 23 km (14 mi) thick layer beneath the clouds and in a layer between 20 and 15 km (12 and 9 mi). There are no rising air currents in these regions, even though high temperatures in the deep atmosphere were expected to provide hot, rising masses of atmosphere that would produce turbulence. Also, before the atmosphere was probed, theoretical work had suggested that the lower atmosphere of Venus would be unstable.

Soviet scientists found that there are, however, some regions of turbulence in the atmosphere above 40 km (25 mi) and this turbulence is greatest at the part of the atmosphere which directly faces the Sun: the subsolar zone. On the night side at the same altitude there is much less turbulence.

The high surface temperatures measured by many probes are approximately the same when corrected to a common distance from the center of Venus. They also correspond with surface temperatures measured from Earth at short radio wavelengths. This very high surface temperature sets Venus apart from Mars and Earth. An objective of probe missions to Venus was to ascertain if the high surface temperature of the planet is a result of the greenhouse effect. As mentioned earlier, results from the U.S. and Soviet probes did confirm that the greenhouse mechanism is responsible.

The probes revealed that the 20-km (12-mi) thick clouds of Venus are in three well-defined, distinct layers, and seem to result from a vigorous cycle of sulfur-hydrogen-oxygen reactions. The clouds appear to be composed mostly of oxygen, water vapor, and sulfur dioxide, and they extend from 70 km (43 mi) to 47.5 km (30 mi) above the mean surface.

From about 30 km (19 mi) down to the surface, the atmosphere appears free of particles, but starting at 13 km (8 mi) altitude, the two nightside probes saw an unexpected glow increasing as the probes descended. Data from mass spectrometers provided evidence for there being various sulfur compounds near the surface. This could imply that the mysterious glow came from "chemical fires" on the surface or in the very hot and dense lower atmosphere near the surface. Such fires might be fueled by reactions involving the sulfur compounds. Alternatively the glow might have originated from heated or electrically charged surfaces of the probes, from lightning flashes in the Venus atmosphere, or from volcanic activity.

The clouds of Venus were an enigma for generations of astronomers. If you look at Venus through a telescope you see the planet completely covered with a bright veil of unchanging, featureless, yellowish clouds. Earth-based observations had not had much success in obtaining data about these clouds. The notable exception was that observations from high-flying aircraft had produced evidence to explain the nature of the cloud droplets. Scientists discovered that these droplets consist of concentrated sulfuric acid.

The spacecraft missions revealed Venus's physical nature in detail. Data from within the clouds were gathered by the Soviet Venera missions, and the Mariner 5 and 10 flyby spacecraft provided information about regions near the cloud tops. Nevertheless, the exact nature of the clouds and their chemistry and composition were virtually unknown before the Pioneer Venus mission.

Earth-based observations generally do not show any features on the clouds at visible or infrared wavelengths. Observations at near ultraviolet wavelengths, however, reveal indistinct features and hint at some horizontal structure, possibly of clouds (figure 5.6). Strangely, the features appeared to circulate around the planet every four days or thereabout. This contrasted remarkably with the surface rotation period of 243 days as determined by radar observations from Earth. The first big breakthrough came from the imaging system carried by Mariner 10. Many images obtained over eight days confirmed the four-day rotation period (figure 5.7), and from these images scientists were able to measure

Figure 5.6. The markings on Venus have now been identified clearly and their changing forms documented as a result of many hundreds of images obtained by Pioneer Venus Orbiter. (NASA/ARC)

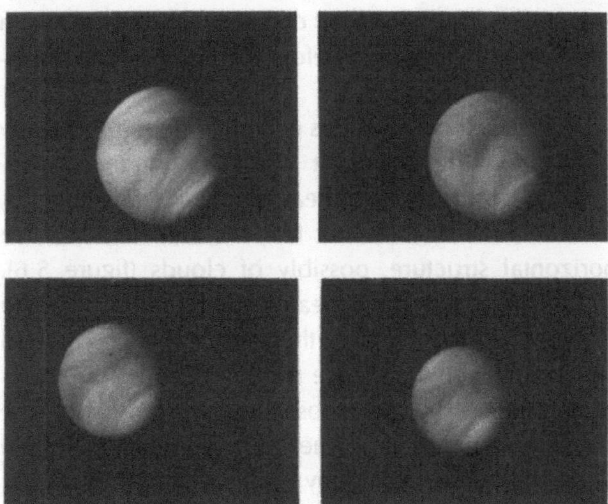

Figure 5.7. The first sequences of images from a spacecraft—Mariner 10—confirmed the four-day rotation.

patterns of circulation near the cloud tops. The images showed that the motions are generally in a direction parallel to the equator, i.e. zonal.

Measurements of sunlight scattered by the uppermost layers of Venus's atmosphere were made from Earth. These included how the polarization of the light changes with variations in the angle of observation of the clouds relative to the solar illumination. From such measurements and from theoretical work, evidence mounted about the detailed properties of the particles of which the uppermost clouds are composed. Best agreement of theory with observations was obtained when the particles were all assumed to be spherical with an effective radius of about 1.05 micrometers and a visible light index of refraction of 1.44. (20 micrometers are approximately equal to 1/1000th of an inch). These conclusions agreed with the interpretation of aircraft-based spectroscopic observations, which suggested that the upper-cloud particles were composed principally of concentrated sulfuric acid.

When the Veneras 9 and 10 penetrated the atmosphere they obtained data to support the conclusions derived from the Earth-based observations. Data from a light-scattering experiment aboard these probes showed that the clouds are in three main layers. Also the Russian probes gathered information about the variations of effective particle sizes and indices of refraction in each of these layers and other regions of the atmosphere which were confirmed by the Pioneer Venus probes.

Since data from these experiments suggested that large particles with large indices of refraction were present, these particles were tentatively identified as large sulfuric acid droplets. In turn, it was suggested that sulfur crystals might be present and partly account for the ultraviolet markings on the top cloud layer.

Pioneer experiments investigated cloud properties throughout the layers and the circulation of features at the tops. Some experiments were aimed at finding how the structure of the clouds varies with altitude at the four locations on the planet where the probes were targeted to enter the atmosphere. While the probe data covered a brief time only, the Orbiter gathered years of cloud-top observations. Like the Russian spacecraft the U.S. probes carried a nephelometer, a spectrometer, and a solar net flux radiometer and an infrared radiometer. The Large Probe also carried a neutral mass spectrometer. The Orbiter had a cloud photopolarimeter/imager, an infrared radiometer, and an ultraviolet spectrometer. Further supporting information was obtained from atmospheric structure experiments by temperature, pressure, and acceleration sensors.

The data provided by measurements within the clouds at the location of each of the four probes, combined with the Orbiter's planetwide observations and similar information from Soviet orbiters and landers, have now given atmospheric physicists, a much better understanding of the clouds, their morphology, their particles, their physical and chemical composition, their optical properties, and their interaction with atmospheric motions.

Cloud regions have been identified (figure 5.8). An upper haze region, from about 70 to 90 km (43 to 56 mi), consists of very small particles. The main cloud deck consists of upper, middle, and lower cloud regions—each with a somewhat different mix of particles. A lower haze, from 47.5 km down to about 31 km (30–19 mi), possibly extends into even lower altitudes. Thin layered structures, which are probably precloud

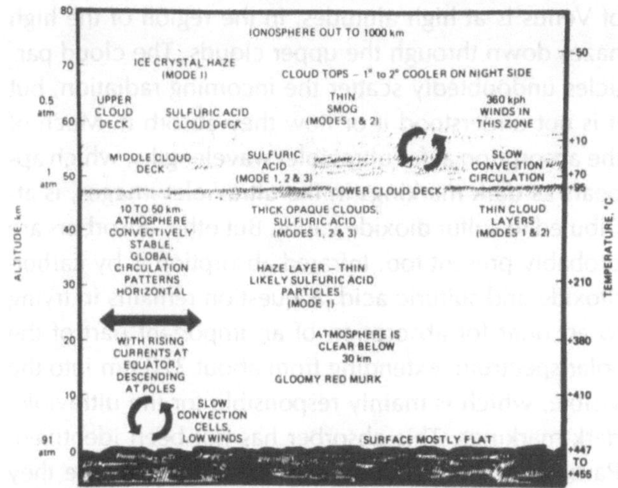

Figure 5.8. Regions of the clouds of Venus have now been identified as a result of the space missions.

layers, exist periodically in the upper part of the lower haze region.

The top cloud layer extends over an altitude range of 70 down to 56 km (43–35 mi). It consists of sulfuric acid droplets of 0.1 to 0.5 and 1.8 to 2.8 micrometers diameter, with about 300 particles per cubic cm. Temperature in the layer was around 13° C (55° F).

The second cloud layer extends down from 56 to 49.5 km (35 to 31 mi). This layer appears to consist of some 0.1- to 0.5-micrometer particles, some 1.8- to 2.8-micrometer particles and some 6- to 9-micrometer particles. At the time of the Pioneer probes' entry the particles were present in concentrations of about 100 per cubic cm. The temperature of the layer was about 20° C (68° F).

The bottom cloud layer is the most opaque. It extends down from 49.5 to 47.5 km (31 to 30 mi) altitude and had 400 particles per cubic centimeter at the time of the Pioneer probes' entry, including many large particles of 6 to 9 micrometer diameter. Its temperature was about 202° C (395° F).

A precloud layer of droplets similar in composition to the top cloud layer—with 300 particles per cubic cm—was present at about 47.5 km (29 mi) altitude. Below was a faint haze of very small particles present in density of about 20 per cubic cm. Haze also extended through the cloud layers and above them. Table 5.1 summarizes properties of the hazes and main clouds as now understood.

The main cloud decks of the upper, middle, and lower cloud regions have particles of various sizes as contrasted with the hazes where the particles are of one size group. On the lower and middle cloud regions there are three sizes of particles with diameters of 0.1 to 0.5, 1.8 to 2.8, and 6 to 9 micrometers. The upper cloud region has two only. The larger particles are absent. The smallest particles comprise a population which extends throughout the main cloud deck and 15 to 20 km

Table 5.1. Characteristic of the Clouds of Venus

Region	Altitude km	Temp. °C	Refr. Indx	Composition	Diameter micrometers
Upper Haze	90.0 – 70.0	−83 to −48	1.45	sulfuric acid + contaminants	0.4
Upper Cloud	70.0 – 56.5	−48 to 13	1.44	sulfuric acid + contaminants	0.4, 2.0 (bimodal)
Middle Cloud	56.5 – 50.5	13 to 72	1.42 1.38	sulfuric acid + contaminants + crystals	0.3, 2.5, 7.0 (trimodal)
Lower Cloud	50.5 – 47.5	72 to 94	1.32	sulfuric acid + contaminants + crystals	0.4, 2.0, 8.0 (trimodal)
Layers	46.0 – 47.5	94 to 105	1.46 1.50	sulfuric acid + contaminants	0.3, 2.0 (bimodal)
Lower Haze	47.5 – 31.0	94 to 209	–	sulfuric acid + contaminants	0.2

(9.3 to 12.4 mi) above and below it. These small particles are sufficient in number to provide centers for growth into larger particles by condensation of atmospheric vapors. The second size of particles consists of sulfuric acid droplets which grow in the upper cloud region and gravitate to lower levels. The sizes correspond to definite layers and the largest appear to be thin plate-like crystals.

The various particle sizes suggest that the particles have several different chemical constituents. Optical properties of the medium-sized particles which are present throughout the main cloud decks are best explained if these particles are sulfuric acid droplets. The concentration of acid in these particles may vary from 90 percent at 60 km (37 mi) to 80 percent at 50 km (31 mi), without affecting the size of the individual drops. The smallest particles have optical properties that indicate a variable composition, but the main component is sulfuric acid in the upper and lower cloud layers, the precloud layers, and the upper haze regions. Sulfuric acid originates above the boundary between the upper and middle clouds. Of what material the largest particles consist is not known. However, the particles may be chlorides because large amounts of chlorine were detected by Venera 11.

The structure of the upper and lower cloud regions varies much more than that of the middle cloud region. All the clouds of Venus are in layers on a fairly large scale. They are also more like fog or mists than turbulent terrestrial clouds. Within them there might be light drizzle rather than heavy rain. Also, the characteristics of the clouds were very similar at all the probes' entry points, suggesting that the clouds possess planetwide rather than local structures.

The observed ultraviolet markings are thought to be associated with the motion of an ultraviolet absorber in the cloud regions. The nature of this absorber and its extent within the various layers is unknown, although some candidates have been suggested. Certainly it seems that changes in the concentration of sulfuric acid particles cannot alone account for the observed patterns.

In contrast to Earth, where the major absorption of solar energy is close to or at the surface in the tropics, the major absorption of solar energy in the atmosphere of Venus is at high altitudes, in the region of the high hazes down through the upper clouds. The cloud particles undoubtedly scatter the incoming radiation, but it is not understood if or how they absorb it. Much of the absorption at far ultraviolet wavelengths, which appears as dark markings in the ultraviolet images, is attributed to sulfur dioxide vapor. But other absorbers are probably present too. Infrared absorption is by carbon dioxide and sulfuric acid. A question remains in trying to account for absorption of an important part of the solar spectrum, extending from about 320 nm into the visible, which is mainly responsible for the ultraviolet dark markings. This absorber has not been identified. Particles of pure sulfuric acid do not fit because they are transparent at the appropriate ultraviolet wavelengths. The missing absorber also appears from photopolarimetry to be deep below the hazes—i.e., in the main cloud decks.

Yet the mystery deepens because some radiometer data gathered by probes indicate that absorption of solar energy occurs at altitudes in or above the upper clouds. The mysterious absorber appears to be located in the same regions as the sulfur dioxide absorber and somehow associated with it.

General features observed at ultraviolet wavelengths (see figure 5.9) may be classified as being associated with three distinct regions of Venus; a polar zone above 50 degrees latitude, a mid-latitude zone between 20 and 50 degrees, and an equatorial zone extending about 20 degrees north and south of the equator. The haze of small particles which completely envelopes the planet differs in density at different latitudes. There is indeed a collar of polar haze (which appears bright in ultraviolet light) encircling the planet at about 55 degrees latitude. Also there is evidence of haze increasing over the morning and evening terminators. At the polar regions features which appear in the infrared images are obscured in the ultraviolet images by the haze. In addition, the general covering of haze appears to vary over times ranging from months to years.

The many different ultraviolet features seen in midlatitudes and equatorial regions basically consists of three major types—bow shapes, dark midlatitude bands, and a dark equatorial band (figure 5.10). The dark equato-

Figure 5.9. The ultraviolet markings have been charted in great detail in images such as these from Pioneer Venus Orbiter. (NASA/ARC)

rial band forms a tail which together with a bow feature produces the horizontal Y-feature which has often been observed from Earth. Sometimes the Y feature is not visible. At other times it retains its structure as it moves around the planet. But many of the feature's detailed characteristics change, thereby suggesting that the much smaller ultraviolet markings change independently of each other. Sometimes the Y is so changed that it forms a C or other shape. In general, the planet displays a whole range of global cloud patterns in addition to the Y pattern.

Atmospheric cells with either dark or bright surroundings are common at low latitudes. Most have dark centers and are about 200 to 300 km (125 to 186 mi) in diameter. They appear in bright and dark regions but are more numerous in the dark equatorial region and during the early afternoon on the planet.

Data received from the infrared radiometer of Pioneer Venus Orbiter (figure 5.11) showed a dark polar band at about 65–75 degrees latitude. It is a broad feature surrounding the polar region and is assumed to be a cold region above which the temperature of the atmosphere increases. The coldest part of the collar seems to be associated with midnight on the planet. These polar collars have been observed from Earth and they usually are associated with one pole during the period when the planet is located favorably for observation. They have sometimes been seen to last for months. During the period when a collar is present, the associated pole also appears brighter than usual.

Spacecraft infrared images have shown that there is a pair of hot spots either side of the pole at about 85 degrees latitude, appearing in the form of a dipole (figure 5.12) which is about 2000 km (1250 mi) long and about 1000 km (620 mi) across. These hot spots are probably the result of clearings in the polar cloud deck that permit observations to penetrate to lower and hotter levels of the atmosphere. The dipole rotates about the planet's pole in about 2.7 days. Bright streaks from one hot spot of the dipole appear sometimes as filaments cutting across the collar. These polar hot spots may indicate that atmosphere is descending in the polar regions, and since hot spots, or descending regions, are not seen anywhere else on the planet, it seems likely that a single large circulation cell fills a hemisphere from the equator to the pole as mentioned earlier.

The cloud system is involved in the general circulation of the atmosphere with high wind velocities and a rapid change in the velocity at the base of the cloud layers. Venus's predominant weather pattern is thus the high-speed circulation of the middle and upper atmosphere around the planet. The motion of the atmosphere is predominently from east to west at speeds up to 360 kph (225 mph). There is also movement of atmosphere at only a few meters per second from the equator to the poles at the cloud tops. As mentioned earlier this is part of a dominant Hadley circulation cell in each hemisphere the return flow of which is at the bottom of the cloud layers. The cell is completed by rising atmospheric masses at the equator and falling masses at the poles, possibly associated with the polar dipole. Other Hadley cells appear to be associated with the main Hadley cell in each hemisphere.

Measurements of wind velocities and temperatures at various levels of the atmosphere by the four probes

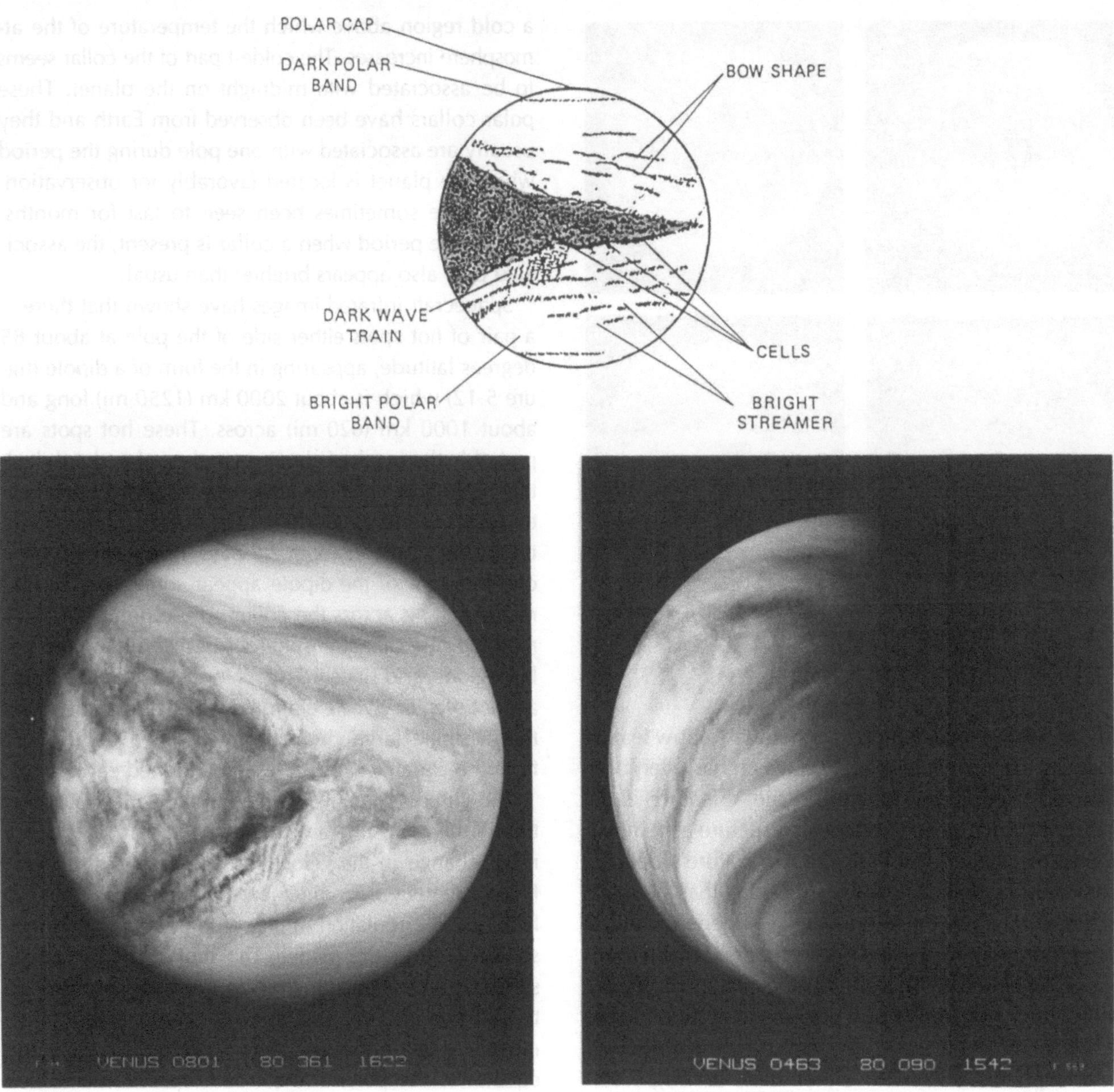

Figure 5.10. Identification of features in the ultraviolet markings of Venus. (NASA/ARC)

of Pioneer Venus indicate that the meridional winds result from a series of equator-to-pole circulation cells, stacked one on top of the other with each of the upper three cells counter-rotating like meshed gear wheels (see figure 5.13). All the winds in these equator-to-pole circulation loops are driven by the solar energy absorbed primarily in Venus's dense, high cloud layer. The whole complex of stacked cells carries Venus's solar heat, absorbed near the equator, to the polar regions. Because the rotation of the planet is so slow, rotation forces do not break up these huge, hemisphere-spanning circulation loops as occurs on the fast-spinning Earth.

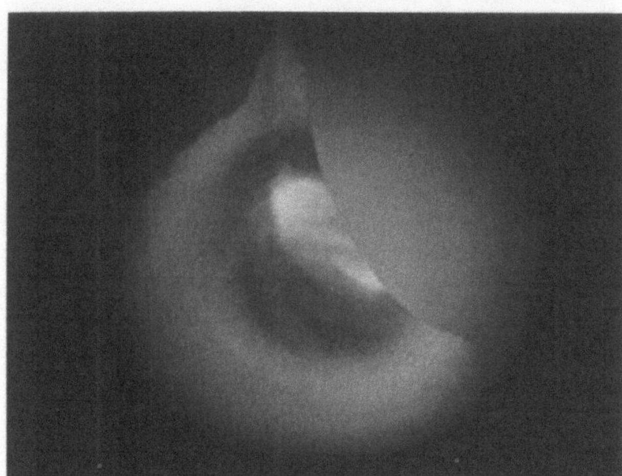

Figure 5.11. Infrared data have been amassed to provide heat maps of the planet. (NASA/ARC)

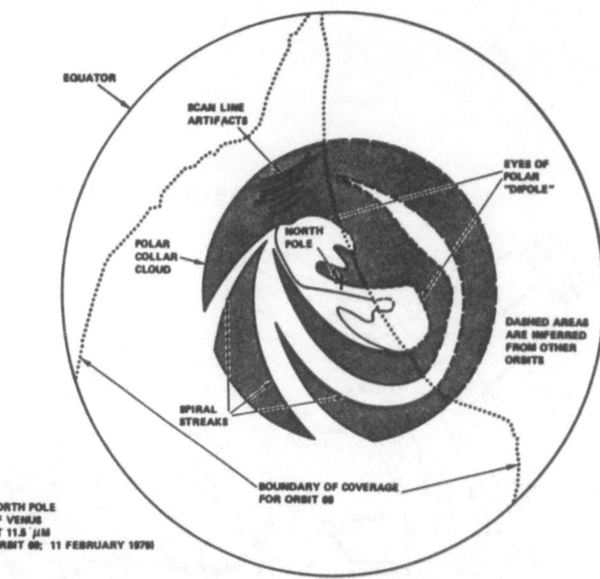

Figure 5.12. The infrared data revealed a polar dipole with spots where atmosphere is descending to lower levels. (NASA/ARC)

As discussed earlier most solar energy is absorbed on Earth at or close to the surface, and the northern and southern Hadley cells are close to the surface and within the troposphere. The cells break up into eddies at mid-latitudes because of Coriolis forces resulting from Earth's rapid rotation. On Venus most of the solar energy is absorbed in the planet's mantle of dense clouds at about 50 km (31 mi) above the surface, and the primary Hadley cells are at altitude. The planet's equator-to-pole circulation is driven by a cell in the cloud regions in the north and south hemisphere. Temperature differences in the cloud cell are relatively high compared with the rest of the atmosphere—a drop of 20° C (36° F) between the equator and 60° latitude. This temperature range and associated pressure differences are comparable with those found at sea level on Earth. Atmospheric pressure at the altitude of the clouds of Venus is, in fact, about the same as that at sea level on Earth.

Above the cloud layer in each hemisphere is probably an upper cell, from 65 to 85 km (39 to 51 mi) altitude. It is not driven by solar heat but by friction with the cloud cell. It accordingly runs in the opposite direction. Below the cloud cell is a lower cell from 45 down to 40 km (27 to 24 mi) which has been called the subcloud cell. It, too, is driven by friction and runs in the opposite way to the cloud cell. Near the surface a fourth cell is believed to operate and this may be driven by the relatively small amount of heat absorbed at the surface of the planet. Between the surface cell and the subcloud cell at 30 to 40 km (18 to 24 mi) altitude there may be small cells which are more akin to very large eddies. Because the atmosphere is stable and not convectively overturning at those altitudes, the eddies may be horizontal instead of vertical.

Because of these two kinds of circulation—around the planet and equator-to-poles—the atmosphere is thoroughly mixed. In turn, as has been observed, it is about the same temperature and pressure everywhere at the same level—except in the high atmosphere above the clouds.

Because the equator-to-pole winds are much slower than those around the planet, winds on Venus blow mostly around the planet but also slowly spiral toward the poles. The combination of these zonal and meridional motions probably leads to the vortices in the polar regions. Those vortices, in turn, affect the haze layer and produce a lowering of the cloud tops in the vortices. These vortices might also cause thickening of high-latitude hazes and the cold collar mentioned earlier.

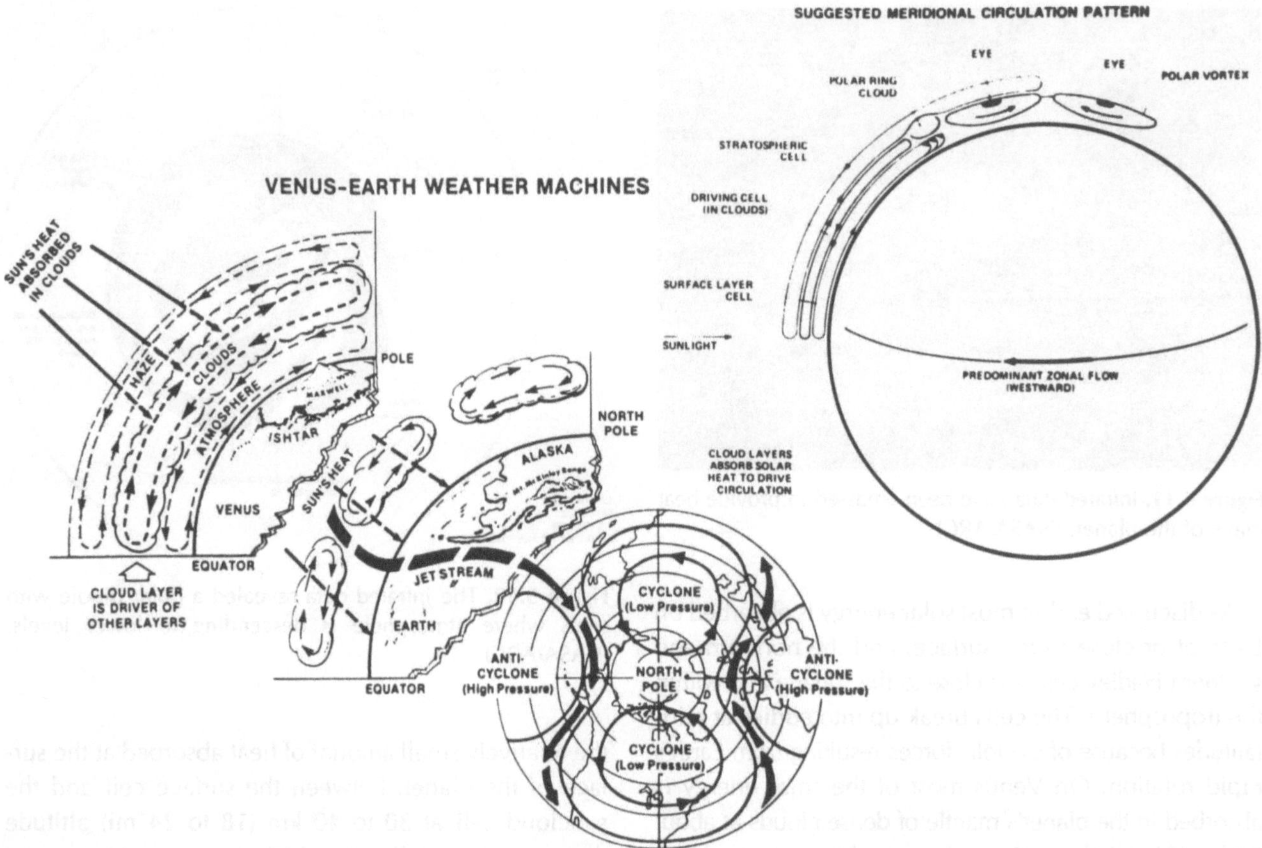

Figure 5.13. North-south circulation moves heat from the equator to the poles through a series of stacked Hadley cells. (NASA/ARC)

In a general way, the high-speed winds can now be explained: when a mass of Venus's atmosphere moves upward because of solar heating, it carries some momentum of the solid planet upward. On successive passes around the Hadley cell, as the atmosphere circulates globally, the momentum accumulates at the upper levels. Some other explanations have also been suggested but the exact mechanism is still not clearly defined.

The growth of cloud particles does not seem to be strongly influenced by the planetary circulation. The acid particles are carried along by the moving atmosphere and their acid concentration adjusts to new equilibriums established at different parts of the circulation cycles. The sulfuric acid droplets appear to grow very slowly except in the lowest cloud regions, where growth could be very quick because of rapid condensation. Particles might last for years in the upper hazes and for only hours in the lower cloud layer.

Measurements by the Pioneer probes showed that Venus's atmosphere is convectively overturning only within the main cloud deck between 53 and 56 km (33 and 35 mi) altitude, and in a layer below the clouds between 20 and 28 km (12 and 17 mi) altitude, and possibly in a third layer between the surface and 6 km (3.6 mi). The rest of the atmosphere appears to be stable. But there is some evidence in these stable layers of sloshing gravity waves driven by the rapid passage of the atmosphere from the dayside to the nightside, with resultant cooling on the nightside and reheating as the atmospheric masses enter the sunlit hemisphere.

Because the atmosphere is so stable from a convective standpoint, it was difficult to reconcile observed electromagnetic signals as being generated by lightning flashes in the atmosphere of Venus. The whistler-type signals were observed by instruments carried by Veneras 11 and 12, and Pioneer Venus Orbiter (figure 5.14). The signals are believed to come from lightning because they are intense and highly impulsive. Pioneer detected them many times when the Orbiter was close to periapsis.

We know that on Earth lightning requires the presence of large particles in the atmosphere coupled with strong updrafts in cloud regions, but there is little evidence of strong updrafts in the clouds of Venus. Also, there is no direct evidence for large precipitative-type particles. If cloud processes are generating the lightning then large undetected particles must be present in the atmosphere. One suggestion is that the lightning is associated with volcanic eruptions or strong convective motions near the midday equator, motions which have not yet been detected in the data.

Attempts to see lightning optically with the Orbiter's star sensor were unsuccessful. But they could look for it only on the dark side, and it is possible (though unlikely) that lightning is confined to the day side of the planet. In a later chapter the association of electrical discharges with volcanic plumes is discussed.

Although our knowledge about the clouds of Venus has been enormously increased by the successful missions to the planet, there are still unanswered questions. The total identity of the ultraviolet absorber or absorbers is still not clear. The composition of the largest particles and the nature of contaminants in other cloud particles is still unknown. The role of chlorine in cloud chemistry is unknown. We are still ignorant of

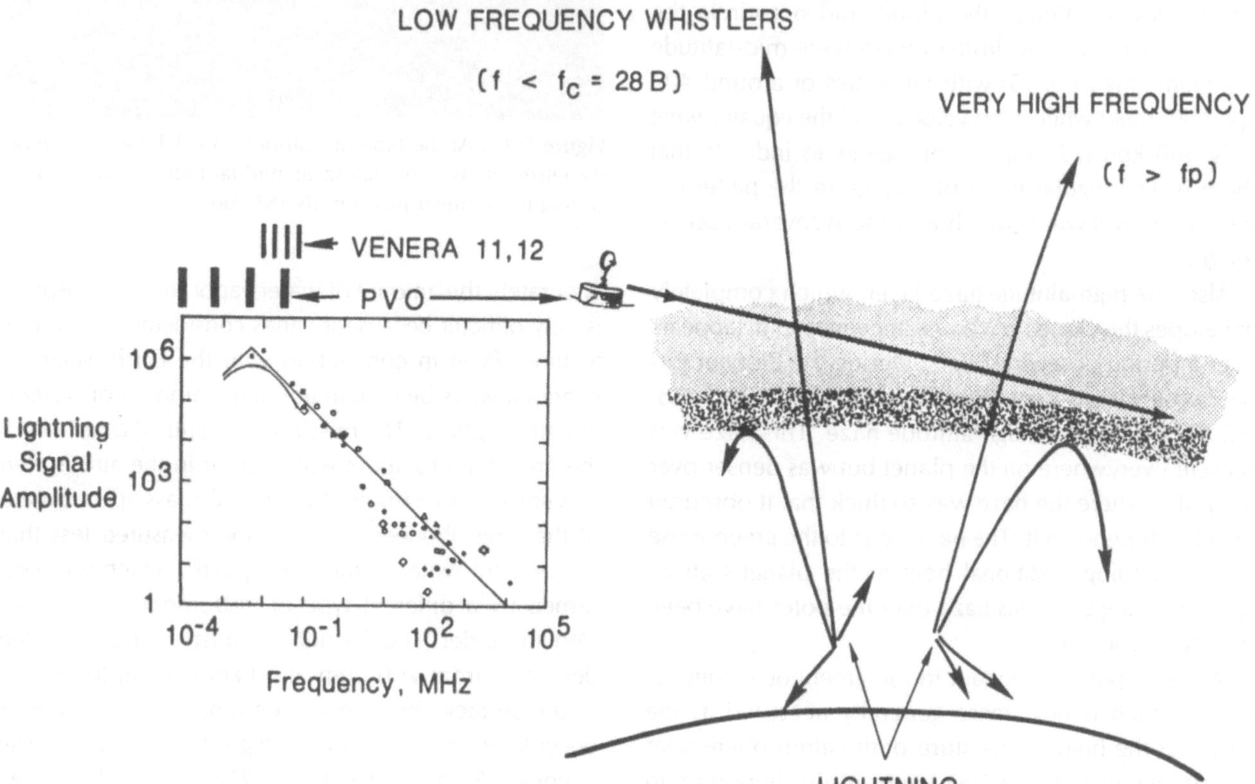

Figure 5.14. Radio signals were detected by Orbiter which suggest the occurrence of lightning flashes within the planet's dense atmosphere. (NASA/ARC)

the nature of the particles suspended in the atmosphere at low altitudes.

Significant discoveries concerning the patterns of atmospheric changes on Venus were made on the basis of two years of observations from Pioneer Orbiter compared with earlier observations by flyby spacecraft and ground-based observations. A long-term change for both the planet's wind patterns and for the presence and absence of the deep haze layer above the clouds is apparent. Venus's planet-wide wind patterns change dramatically over a few years. Two circulation patterns have been recognized in which a mid-latitude jetstream was succeeded after several years by a pattern of cloud and wind circulation like that of a single rigid body—i.e., with velocity decreasing with increasing latitude.

The solid-body rotation of cloud patterns, with winds moving faster at the equator than closer to the poles, applied during two years of the Venus Orbiter observations. But several years earlier, in 1974, when Mariner 10 flew by Venus, the clouds did not circle the planet as a rigid body. Instead there were mid-latitude jetstreams (figure 5.15) with velocities of around 400 kph (245 mph) while wind velocities at the equator were only 360 kph (220 mph). This seems to indicate that there is an irregular cycle of change in the pattern of the cloud level winds which is perhaps several years in length.

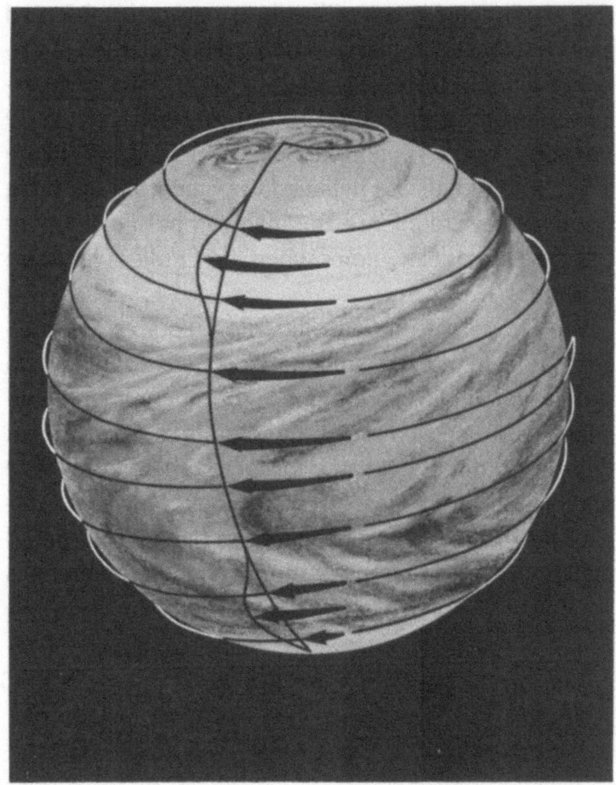

Figure 5.15. At the time of Mariner 10's flyby a high-velocity jetstream was operating at mid-latitudes. It was absent during the Pioneer mission. (NASA/ARC)

Also, the high-altitude haze layer, which completely envelopes the clouds of Venus, appears and disappears over a period of several years. During the Pioneer Orbiter observations Venus was enveloped in a 29-km (18-mi) thick blanket of high-altitude haze. This haze was present everywhere on the planet but was denser over the poles. There the haze was so thick that it obscured the clouds beneath it. The haze adds to the greenhouse effect, trapping additional heat in the planet's atmosphere. Changes to this haze over the poles have been observed from Earth.

Water vapor is important in the greenhouse mechanism, which is now more generally accepted as the cause of the high temperature of the atmosphere near the surface of Venus. Water vapor is also important to the chemical composition of the atmosphere and the reactions in the atmosphere. Unfortunately, measuring accurately the amount of water vapor in an atmosphere is very difficult because it varies considerably from time to time. Even in connection with the Earth, scientists cannot always be certain about the amount of water in our stratosphere. The missions to Venus did not clarify the uncertainties about water vapor in the atmosphere of Venus. For example, the neutral mass spectrometer of the large Pioneer Venus probe measured less than 0.1 percent water in the atmosphere, which was confirmed by a different type of instrument carried by a Venera lander. The Russian scientists found that water decreases from 200 ppm at 50 km (31 mi) to 20 ppm at the surface. But the gas chromatograph of the Pioneer Venus probe found much greater amounts of water vapor—0.52 percent at 42 km (26 mi) and 0.13 percent at 22 km (13.5 mi).

The quantity of carbon monoxide in the atmosphere

of Venus is extremely small also—about 20 ppm at 22 km (13.5 mi). Measurements from Earth indicate that it is about 50 ppm at the cloud tops. Such a distribution would be expected if carbon monoxide is produced above the clouds by photodissociation of carbon dioxide and from there diffuses downward. But the amount of carbon monoxide expected to be vented at the same time as carbon dioxide from the interior of Venus would be at least one thousand times the amount observed. Carbon monoxide can react with water to form hydrogen and carbon dioxide early in the history of the planet, and this could have taken place on Venus. If so, it is a possible explanation for the lack of water on the planet today: the hydrogen would have escaped into space. It seems most unlikely, however, that the initial amounts of water and carbon monoxide would be sufficiently right for them to react together to leave such minuscule quantities of each as are now measured on the planet.

Oxygen has been detected by various instruments, increasing from 16 ppm to 43 ppm between 42 and 52 km (26 and 32 mi) as contrasted with 1 ppm measured at the cloud tops from Earth. How carbon monoxide and molecular oxygen can exist at the same time in the atmosphere is difficult to explain. Photolysis of carbon dioxide above the clouds produces oxygen and carbon monoxide, the amounts present below 52 km (32 mi) are inconsistent with the small quantity above the clouds. There is no generally accepted explanation for the presence of this oxygen.

The various probes show that carbonyl sulfide is present in amounts less than 3 ppm, but sulfur dioxide at 22 km (13.6 mi) is present in fairly large amounts—130 to 185 ppm—decreasing to an almost negligible amount (0.1 ppm) above the clouds. Hydrogen sulfide gas is present in the amounts of about 3 ppm at the surface to 1 ppm in the clouds. As discussed earlier, the clouds contain large amounts of sulfuric acid. Before Pioneer Venus a cycle of chemical reactions similar to one responsible for formation of sulphate aerosol layers on Earth was proposed for Venus. Carbonyl sulfide played a key role in this scheme, but it was found only in very small amounts. So atmospheric chemical cycles using a sulfur dioxide and water source to produce sulfuric acid are now favored.

The water vapor measurements present major problems to theoreticians. Use of the high value obtained from the Pioneer Venus gas chromatograph in a thermodynamic calculation predicts hydrogen sulfide and carbonyl sulfide concentrations somewhat larger than those derived from the gas chromatograph results. They are, however, consistent with measurements made by the mass spectrometer. The smaller concentration seen by the Venera photometer would not allow as much hydrogen sulfide as found by the mass spectrometer. Whether the gas chromatograph measurement of 0.52 percent water at 52 km (32 mi) or the photometer value of 200 ppm is correct, compounds containing equivalent amounts of hydrogen atoms would be expected to exist at the surface; but there are no reports of such hydrogen compounds being detected by the Venera landers.

Important discoveries have been accumulated about the solar wind and how it interacts with Venus and its ionosphere. The ionosphere of a planet is defined as the region of the upper atmosphere where there is a large population of electrons and ions in the atmospheric gases. These electrically charged particles are produced by solar radiation in the extreme ultraviolet region of the spectrum, interacting with uncharged (neutral) molecules and atoms of the upper atmosphere. The types and densities of ions found in an ionosphere depend on the general composition, the chemical reactions that occur, and how the ions are transported from place to place within the ionosphere. The behavior of a gas consisting of charged particles (known as a plasma) is affected by magnetic fields.

By measuring how radio signals are delayed in their passage from a spacecraft to Earth, scientists can ascertain the number of electrons (electron density) encountered along the path of the signal. If the radio waves pass through the atmosphere of a planet on their way to Earth, information can be obtained about the ionosphere. By directing the spacecraft along a path that carries it behind the planet as viewed from Earth, an occultation occurs and the radio waves travel through the atmosphere as the spacecraft moves behind or emerges from behind the planet. In this way early missions to Venus provided information on the total electron densities in the planet's ionosphere. The Pioneer Venus Orbiter experienced many occultations during the mission, as did various Soviet orbiters. The orbiter also made measurements within the ionosphere during periapsis passages.

With measurements made by these spacecraft, scientists found that the ionospheric electron density reaches a maximum at about 145 km (90 mi) on the day and the night hemispheres. Above this altitude the electron density decreases gradually with increasing height. Many unusual ionospheric phenomena were discovered, including ionospheric density depletions—so-called holes—and detached plasma clouds.

Pioneer discovered that the solar wind interacts with the atmosphere of Venus several times more strongly than expected. Three key regions of the upper atmosphere of Venus almost coincide; the turbopause (where atmospheric mixing begins), the region of the ionosphere having maximum ion density, and the base of the exosphere (where gases escape from the atmosphere into space).

During the mission of Pioneer Venus Orbiter, the planet's bow shock wave in the solar wind was very strong, with powerful upstream plasma waves in front of it, out to a distance of several planetary diameters (figure 5.16). Since Venus has little or no intrinsic magnetic field, the solar wind is not pushed far away from the planet as it is by Earth's magnetic field. Instead, it interacts directly with the top of Venus's atmosphere and hence with the high ionosphere of Venus. Normally the blunt nose of Earth's bow shock is 65,000 km (40,000 mi) away from the planet. But Venus's bow shock was eight times closer; at 8000 km (5000 mi) during the early part of Pioneer's mission. The top of Venus's ionosphere appeared to average about 400 km (240 mi) from the surface of the planet, while the region of the atmosphere where a spacecraft begins to experience drag effects began at 250 km (155 mi).

The region between the bow shock and the top of the ionosphere is about 7500 km (4650 mi) wide at the nose of the shock. But the solar plasma and the magnetic field in this region were turbulent and the temperature of the plasma was about 1 million deg. C (1.8 million deg. F), which is many times the temperature

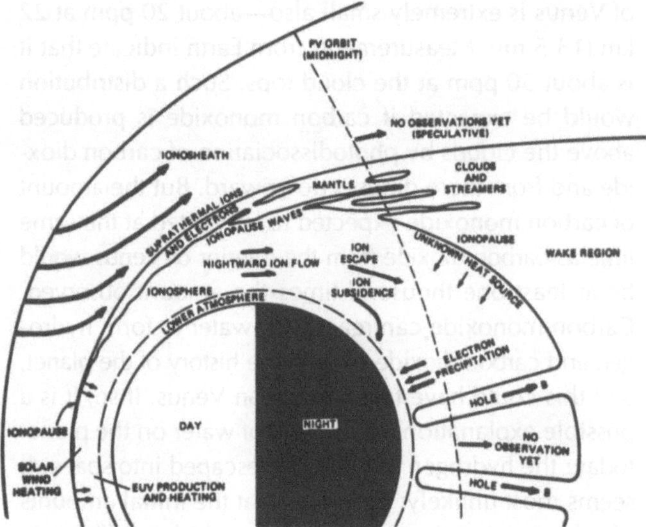

Figure 5.16. The missions to Venus have now built up a good picture of the complex interactions between the solar wind and the planet, the position of the bow shock, and the various regions of the near-space environment of the planet. These are all very different from conditions at Earth. (NASA/ARC)

in similar regions of the near-space environment of Earth. Also there were relatively strong magnetic fields measured at the top of the ionosphere. Because there is no strong magnetic barrier to the solar wind, the ionosphere of Venus is very responsive to changes in solar wind pressure; increases in the solar wind pressure push the ionosphere down toward the planet's surface.

Before the Pioneer results were available there was some speculation that the ionosphere alone could not hold off the solar wind. Unexpectedly, the spacecraft missions discovered that the solar wind is held off Venus as strongly by that planet's ionosphere as it is by Earth's magnetosphere. Pioneer confirmed that the magnetic field of the solar wind streaming past Venus does indeed induce a field in the planet's ionosphere and this field, in turn, holds off the solar wind very effectively, although it penetrates much closer to Venus than it does to Earth.

Spacecraft have discovered other important differences between Earth and Venus. Earth's ionosphere extends to many thousands of kilometers, gradually tapering off with increasing altitude, made possible because Earth's ionosphere is shielded from the solar wind by a strong intrinsic magnetic field. Venus, however—with virtually no intrinsic magnetic field—suffers the consequence that the solar wind can interact directly with the ionosphere. The ionosphere forms an obstacle to the solar wind and deflects it around Venus. Consequently the ionosphere ends abruptly at an altitude of only a few hundred kilometers, which, however, varies considerably depending on the strength of the solar wind. The boundary where the ionosphere ends and the region of decelerated solar wind ionosheath begins is called the ionopause (see figure 5.16). At Venus the solar wind confines the ionosphere below this well-defined ionopause boundary, which changes in step with changes in intensity of the solar wind. The Earth has no such boundary.

The first week of Pioneer's mission was at a period when the speed of the solar wind changed from 500 to 250 km/sec (310 to 155 mi/sec). In this same period the ionopause rose from 250 to over 1500 km (155 to 930 mi) above the surface of Venus. At one point during the mission a solar flare produced a solar wind speed of 600 km/sec (373 mi/sec) and a tenfold increase of pressure on the ionosphere. The result was that the ionopause was pushed down to 250 km (155 mi).

The electron temperature below the ionopause had been expected to be about 730° C (1300° F). The Orbiter found the temperature was much greater—4700° C (8500° F). This was probably caused by some solar wind ions penetrating the ionopause and heating its lower regions.

Just outside the ionopause is a region where a large horizontal magnetic field contains some ionosheath plasma and some rapidly moving "superthermal" plasma, which originated from the ionosphere. This magnetic field, induced by the interaction of the solar wind with the ionosphere, transmits the solar wind pressure to the ionosphere. As mentioned earlier, when this pressure is high the magnetic field is enhanced, and the ionosphere is pushed to a lower altitude. When the solar wind pressure falls, the ionosphere moves up. As a result the ionosphere height varies from 200 km (125 mi) to over 1000 km (625 mi) on the dayside. On the nightside, where there is no direct action of the solar wind, the height of the ionopause is usually less than 1000 km (625 mi).

The magnetic field within the ionosphere of Venus is weak except for unique magnetic structures which have been called flux ropes. They are long, narrow, ropelike regions of strong magnetic field in which the field lines are twisted (figure 5.17). These ropes may be formed from the large magnetic field piled up just outside the ionopause and then drawn down into and through the ionosphere by the solar wind "pulling" on the "ends" of the ropes. Another possibility is that the magnetic flux ropes are generated in a region of large ionospheric magnetic field near the subsolar point.

On the nightside of Venus the ionospheric magnetic field is larger and more regular than on the dayside. It conforms to a symmetrical shape expected to result from

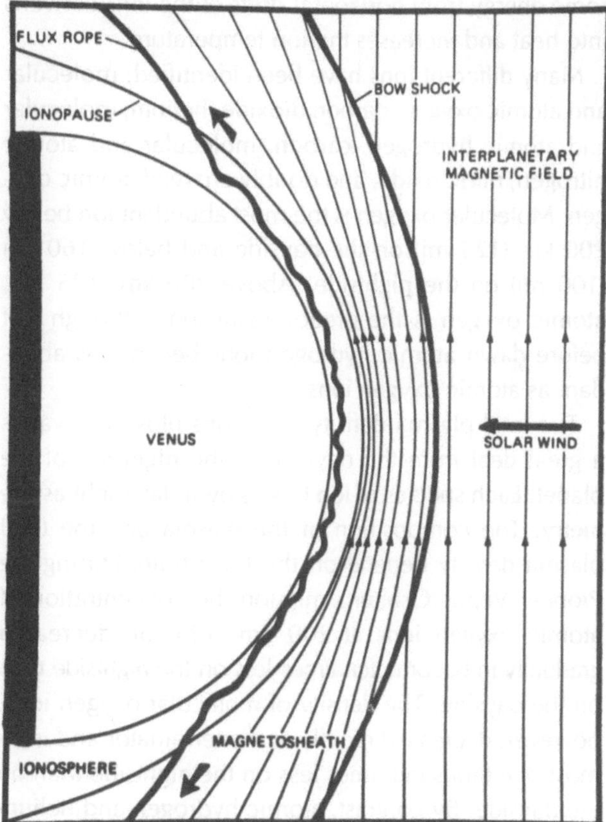

Figure 5.17. Strong parts of the magnetic field near Venus can be likened to magnetic ropes pulled by the solar wind. (NASA/ARC)

a "draping" of the solar wind's field lines around the planet.

Since the movement of electrically charged particles and heat in a plasma is along magnetic field lines, the flux ropes are important in controlling the temperatures of electrons and ions in the ionosphere. The electron temperature is a few thousand Kelvin on the dayside and nightside. This is much hotter than the neutral gas in the thermosphere, which has a temperature of only a few hundred Kelvin. In addition to the effect of the flux ropes another cause of high temperature is heat from the solar wind being "pumped" into the ionospheric electrons at the ionopause. The ions are also at high temperature; about 2000 K on the dayside and more than 4000 K on the nightside. Also friction between the neutral gas and the ions generates heat to raise the temperature of the ions, and on the nightside some energy from horizontal drifts of the ions converts into heat and increases the ion temperature.

Many different ions have been identified; molecular and atomic oxygen, carbon dioxide, helium, molecular and atomic hydrogen, carbon, molecular and atomic nitrogen, nitric oxide, and doubly charged atomic oxygen. Molecular oxygen is the most abundant ion below 200 km (125 mi) on the dayside and below 160 km (100 mi) on the nightside. Above 200 km (125 mi), atomic oxygen is the predominant ion, although just before dawn atomic hydrogen ions become as abundant as atomic oxygen ions.

The total plasma density (electrons plus ions) varies a great deal from the dayside to the nightside of the planet. Each species of ion has its own day/night asymmetry. The composition of the plasma and the total plasma density depend on the local time. During the Pioneer Venus Orbiter's mission the concentration of atomic oxygen ions at 200 km (125 mi) decreased gradually to become ten times less on the nightside than on the dayside. The density of molecular oxygen ions, however, decreased rapidly at the terminator and is almost one thousand times less on the nightside than on the dayside. By contrast, atomic hydrogen and helium ions are more numerous on the nightside than on the dayside, but with more hydrogen than helium ions in the region before dawn.

With the night on Venus lasting for 58 Earth-days—much longer than the lifetime of the ions forming the ionosphere—a question scientists asked was how the nightside ionosphere was maintained. Pioneer Venus Orbiter scientists identified two sources of ionization for the nightside. One source, supported also by data from Venera spacecraft, is the bombardment of the nightside atmosphere by fast electrons. This is similar to the electron flux which produces terrestrial auroras. The energetic electrons originate beyond the ionopause in the wake of Venus. Their bombardment ionizes the neutral gas and suffices to maintain most of the observed ionization in the lower part of the nightside ionosphere.

The other source of ionization is horizontal flows or drifts of plasma from day to night hemisphere at up to 10 km/sec (6.2 mi/sec which can maintain the nighttime ionosphere at high altitudes. This flow also supplements the energetic electron mechanism in maintaining the lower ionosphere; ions flowing to the nightside also fall to lower altitudes.

Day-to-night winds blowing in the neutral atmosphere drag ions along to the nightside. At altitudes near to the ionopause, the flow of plasma away from the Sun induces a flow of charged particles in the ionosphere. At intermediate altitudes the gradients in the ion densities between day and night cause drifts of ions.

The ionosphere varies considerably with time. Pioneer Orbiter measured pronounced fluctuations of ion concentrations from orbit to orbit. The nightside ionosphere sometimes would disappear completely; at other times, it was normal except for strange "holes" in the plasma. At these localities the electron density was very low and the electron temperature very high. The magnetic field in these holes was vertical, which suggests that the holes are associated with the large-scale structure of the field on the nightside. Also, Pioneer Orbiter observed "clouds" of plasma in both night and day hemispheres. These detached layers of plasma lay outside the ionosphere beyond the ionopause, usually close to the terminator. One explanation for the clouds is that the solar wind or flow in the ionosheath pulls plasma from the ionopause region of the ionosphere and carries it downstream.

Magnetic fields of planets are believed to result from a dynamo effect. However, the dynamo theory did not predict the magnetic field discovered by the Mariner 10 spacecraft when it flew by slow-spinning Mercury. Dynamo theories predict that a planetary dynamo, such as the one generating the field of Earth, should depend on spin rate. Other planets visited by spacecraft (Mars, Jupiter, and Saturn) were found to have such fields. If Venus had an internal dynamo identical to Earth's, but weaker in proportion to the spin rate, the planet would have a magnetic field that would be easily detectable. But no such field was found when spacecraft flew by and later orbited Venus.

Except for Venus, every planet seems to have an internally driven magnetic field. Before Pioneer Venus reached the planet there was speculation that Venus might have an internal magnetic field, but one so weak that it could not have been detected by earlier Mariner missions. Pioneer Venus Orbiter ended such speculations. Its highly sensitive instruments detected no intrinsic magnetic field. A major mystery of geophysics is the exact nature of the internal dynamo which generates the Earth's magnetic field and those of other planets. If Venus had possessed an intrinsic field its measurement might have thrown more light on the role played by planetary rotation in the dynamo process.

Current theory states that a planetary magnetic dynamo requires a highly electrically conducting fluid core. The absence of a conducting core may explain why the Moon lacks an intrinsic magnetic field, but it does not explain why Venus does not have one. Temperatures and pressures within Venus are sufficient to allow a highly conducting fluid in the planet's interior. It could be that the composition and electrical conductivity of the fluid is different from those of Earth. Although Venus appears to be Earth's twin in size, it may not be a twin in chemical composition. This would be expected if Venus formed at a different place in the solar nebula and at a different temperature.

At one time it was thought that the interior of the Earth derived heat from decay of radioactive elements. But some of the internal heat may derive from a growing core which releases the latent heat of fusion. This energy source may be stronger than the radioactive heating mechanism.

Pressure and probably temperature at the core of Venus are only slightly less than at Earth's core. However, this difference may be sufficient to prevent solidification of Venus's inner core, even if the two planets have the same internal composition. So the core of Venus may not be growing and adding to the internal heat of the planet.

The surface of Venus and the implications to geology and geophysics and the evolution of Venus are discussed in the next two chapters.

6

VENUS UNVEILED

Highlights of the recent findings about Venus included the obtaining of radar images of the surface from ground-based observations and from spacecraft, and radar altimetry from Pioneer Orbiter for nearly all the surface of the planet. The coverage by radar from Earth was extended by the spacecraft and, coupled with radar images obtained later by the Veneras, the final veils were removed from Earth's errant twin. On the radar images volcanic and tectonic features such as rift valleys, mountains, continents, and volcanoes are apparent. The extensive mapping by Pioneer showed that there is a unimodal distribution of topography (quite unlike the bimodal distribution of continents and deep ocean basins on Earth) and a lack of elevated regions of continental size. The existence of great troughs which may be rift valleys was another important discovery by radar, and later there seemed to be evidence of continuous ridge systems which are characteristic of the terrestrial plate tectonics system.

Concerning the figure of the planet it became clear from measurements of the gravity field combined with the radar altimetry results that the interior behavior of Venus is more like that of Earth than that of Mars or the Moon. There is, nevertheless, a great difference between the two planets: Venus has a strong positive correlation of gravity anomalies with topography.

The inner planets of the Solar System have developed crusts with a composition different from the bulk composition of each planet and from the interior of each planet. The crusts themselves also differ among the terrestrial planets so far explored. How a crust develops depends upon the process of differentiation during which the planet heats to the point at which heavier materials can sink toward the center and lighter materials move upward toward or to the surface. At one time it was thought that the crusts were formed of material that the planet captured by accretion in the concluding phases of its formation. This crustal accretion theory no longer holds vogue. Instead, the crust is believed to consist of material that was accreted in common with other material at the initial formation of the planet and subsequently moved to the surface after the planet heated either during the concluding phases of formation or later because of radioactive heating.

Earth has two distinct crustal types; oceanic crust and continental crust. The young oceanic crust continually

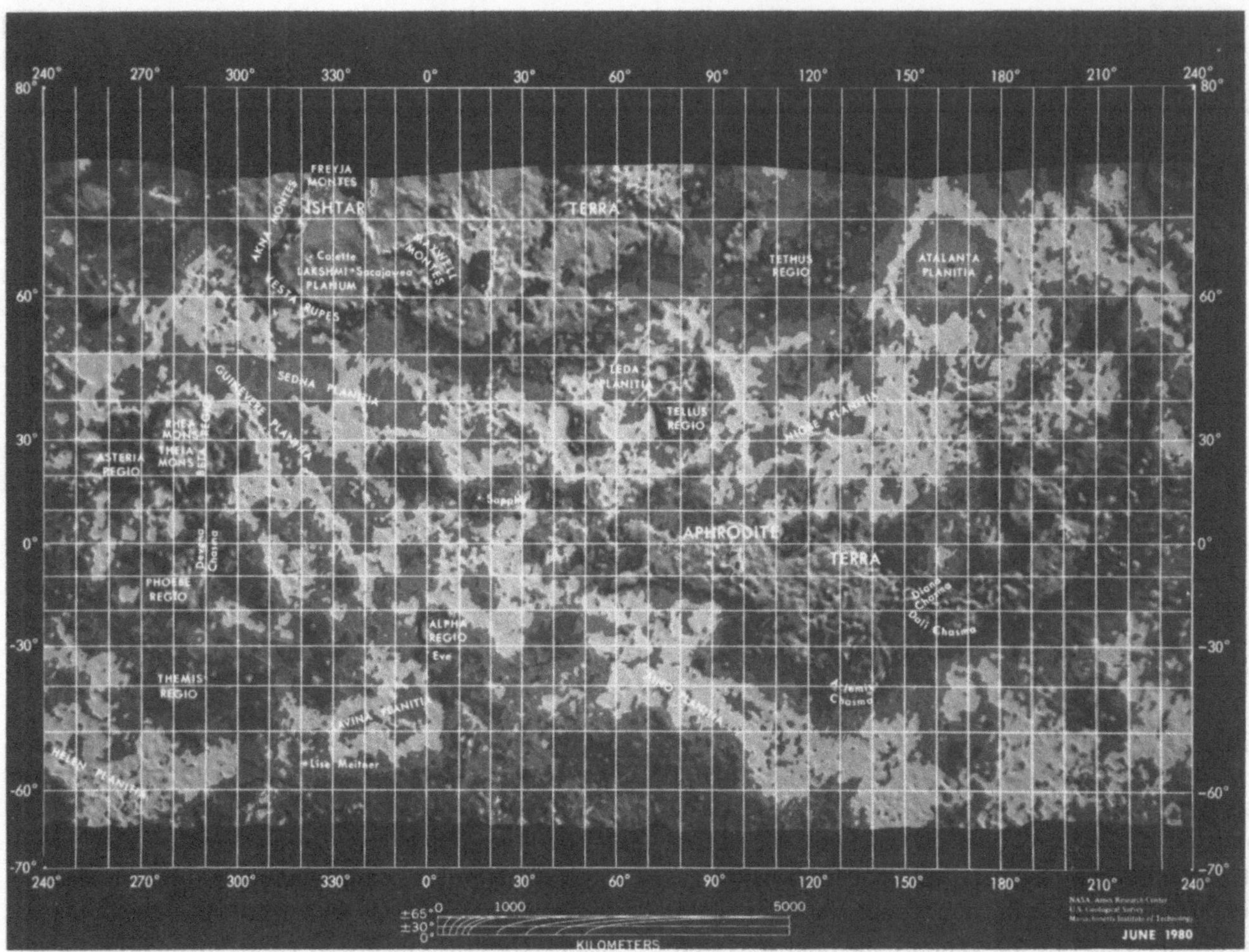

Figure 6.1. (a) This contour map of Venus covers about 83 percent of the planet's surface. The topography consists of highlands, lowlands, and a planet-encircling rolling plain which covers about 60 percent of the surface. (b) (on facing page) identifies the highland areas. (NASA/ARC)

grows from oceanic spreading centers and consists primarily of basaltic material. Oceanic crust is lost by subduction in which it plunges underneath the continents and back into the mantle to be recycled. Continental crust appears to be much older and more complex than the oceanic crust. Moreover it appears to accumulate over geologic time. Some of the continental crust is nearly 4 billion years old; and this type of crust probably formed initially between 4 and 3 billion years ago when the mantle of Earth was much hotter than it is now. Because large amounts of radioactive elements were brought to the surface and incorporated in the crust, further episodes of melting occurred and obliterated most traces of the bombardment from space which is so evident on the Moon, Mars, and Mercury.

By contrast the crust of the Moon originated when the Moon underwent a planetwide episode of melting, followed later by the molding of the surface by bombardment from space and later still by lava flows which occurred predominantly on the hemisphere facing Earth.

Mercury appears to have formed its crust relatively early also, so that it, too, like our Moon, bears the record of the bombardment. Mars, by contrast, evolved its crust over a longer period, with volcanic flows continuing until what appears to be comparatively recent times and leaving smaller areas of cratered ancient crust.

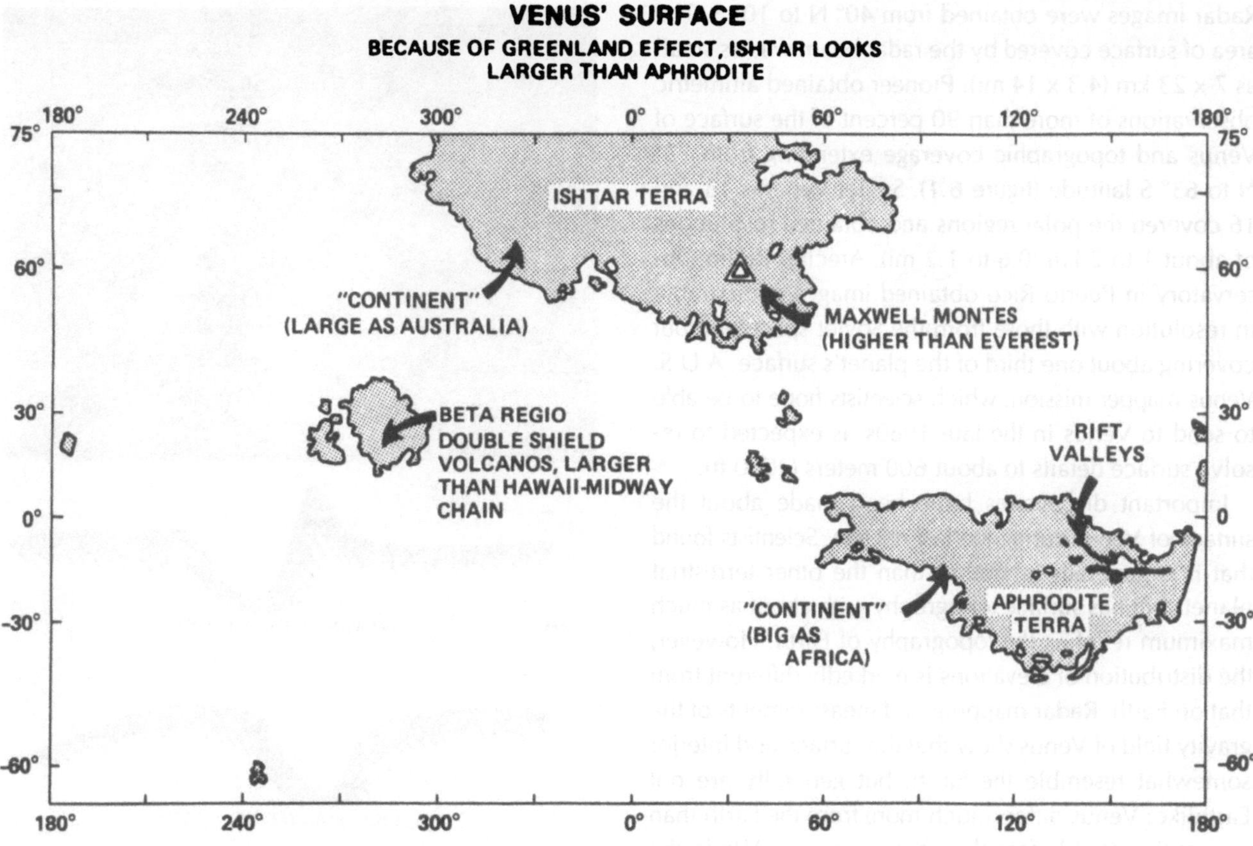

Venus has two continent-sized masses which may indicate continent formation like the continents of Earth. Of particular importance in deciding what happened on Venus is the origin of some circular features on the extensive plains of the planet; are they impact craters or volcanic calderas? If they are impact craters then it means these plains are very old crust, and plate tectonics are unlikely to be occurring on Venus today. Generally it appears likely that the surface of Venus is covered by an ancient crust because the high surface temperature would make subduction difficult on Venus, thereby preventing crust from being recycled into the mantle and stopping further spreading of such crust.

As for the planet in general, radar data provided the first global elevation survey of the surface of Venus from which about 90 percent of the planet was mapped topographically. Before the Pioneer Venus mission the surface of Venus was the least known of all the terrestrial planets. Some radar images had been obtained from Earth, but there are limitations to the coverage by Earth-based radar. Because Venus rotates so that it turns almost the same hemisphere toward Earth when it is closest to us and suitable for radar imaging, Earth-based radar can look in detail at less than half of the planet's surface and along only a narrow equatorial swath. By contrast a spacecraft can travel in a highly eccentric orbit which allows it to cover most of the planet's surface. Pioneer Orbiter, with a 24-hour period and a large orbital inclination, covered most of the planet except for the polar regions, which were later mapped by Venera spacecraft.

Pioneer Orbiter's altimeter mapping sequences were made over a period of about one hour each orbit at altitudes below 4700 km (2920 mi). The Pioneer Venus altimetry radar shows features that are larger than about 75 km (46 mi) in diameter. As the orbit precessed around the planet, the radar view gradually covered nearly all the surface but with lower resolution at high latitudes.

Radar images were obtained from 40° N to 10° S. The area of surface covered by the radar beam was as small as 7 x 23 km (4.3 x 14 mi). Pioneer obtained altimetric observations of more than 90 percent of the surface of Venus and topographic coverage extending from 73° N to 63° S latitude (figure 6.1). Soviet Veneras 15 and 16 covered the polar regions and obtained resolutions of about 1 to 2 km (0.6 to 1.2 mi). Arecibo Radio Observatory in Puerto Rico obtained images comparable in resolution with those from the Soviet spacecraft but covering about one third of the planet's surface. A U.S. Venus mapper mission, which scientists hope to be able to send to Venus in the late 1980s, is expected to resolve surface details to about 600 meters (2000 ft).

Important discoveries have been made about the surface of Venus during the last decade. Scientists found that it is generally smoother than the other terrestrial planets but has surface topography with about as much maximum relief as the topography of Earth. However, the distribution of elevations is markedly different from that on Earth. Radar mapping and measurements of the gravity field of Venus show that the surface and interior somewhat resemble the Earth, but generally are not Earthlike; Venus differs much more from the Earth than was anticipated before the space missions. While the planet, like Earth, probably differentiated into a crust, mantle, and core, it seems to have evolved very differently from a tectonic standpoint and this has greatly affected the topography.

The first result of altimetry was to confirm that the planet is quite round—very different from the other planets and from the Moon. Earth, for example, is flattened at the poles and bulges 21 km (13 mi) at the equator. The Moon has a bulge toward the Earth. Mars bulges, too; but Venus has neither polar flattening nor an equatorial bulge.

Earth has major variations between continents and ocean basins, which cover 30 and 70 percent of the surface respectively. The mean levels of the terrestrial continents and the ocean floors are separated by 4.5 km (2.8 mi). Mars also has major variations and the colossal uplift of the Tharsis region. By contrast, Venus has a very narrow distribution of surface elevations: 20 percent of the planet lies within 125 meters (410) ft) of the mean radius of 6051 km (3760 mi), and 60 percent

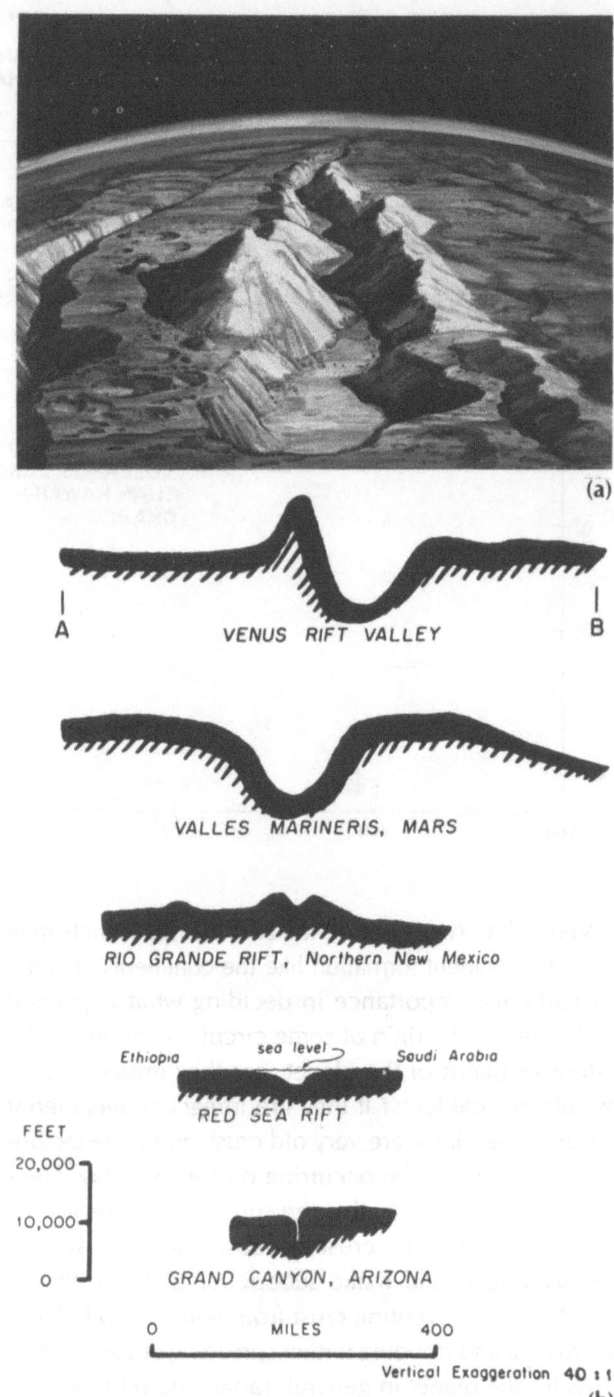

Figure 6.2. A great rift valley was discovered on Venus east of Aphrodite Terra. (a) This artist's concept includes some of the lowest known parts of the planet's surface at the floor of the rift valley. (b) Comparison of the great rift valley on Venus and valleys on other planets. (NASA/ARC)

lies within 500 meters (1640 ft) of it. The planet is a monotonous, dry, hot, and probably dusty world of low relief with only a few large continents and smaller mountainous areas rising above a global plain.

The highest point on Venus is a summit in Maxwell Montes; 10.8 km (6.7 mi) above the mean level and 1525 meters (5000 ft) higher than Mount Everest rises above sea level on Earth. Arecibo images show this area has long, detailed structures that could be folded mountains. They also reveal a large crater or caldera which suggests that Maxwell is an ancient volcano. The lowest known point on Venus is 2.9 km (1.8 mi) below the mean level; in a rift valley located at 204° longitude and 14° S latitude. This depth is similar to but shallower than that of the Vallis Marineris on Mars (figure 6.2), and only one-fifth the greatest depth on Earth—the Marianas Trench.

Comparisons between Earth-based radar observations of Venus and the Moon show that on a small scale Venus is significantly smoother than the Moon at all radar wavelengths. However, one part of Venus is rougher than the roughest area on the Moon.

Estimates of the harmonic coefficients of the gravity field of Venus were computed from long periodic changes of the mean orbital elements of the Pioneer Orbiter. The oblateness of the planet is exceedingly small, as was expected from its slow rate of rotation. Detailed gravity measurements have been made over a significant area of Venus and many anomalies were detected (figure 6.3). But unlike gravity anomalies on the Moon and Mars, their amplitudes are relatively mild and similar to those of Earth. However, unlike Earth the gravity anomalies of Venus correlate with topography. Significant adjustment to the crust of Venus seems to

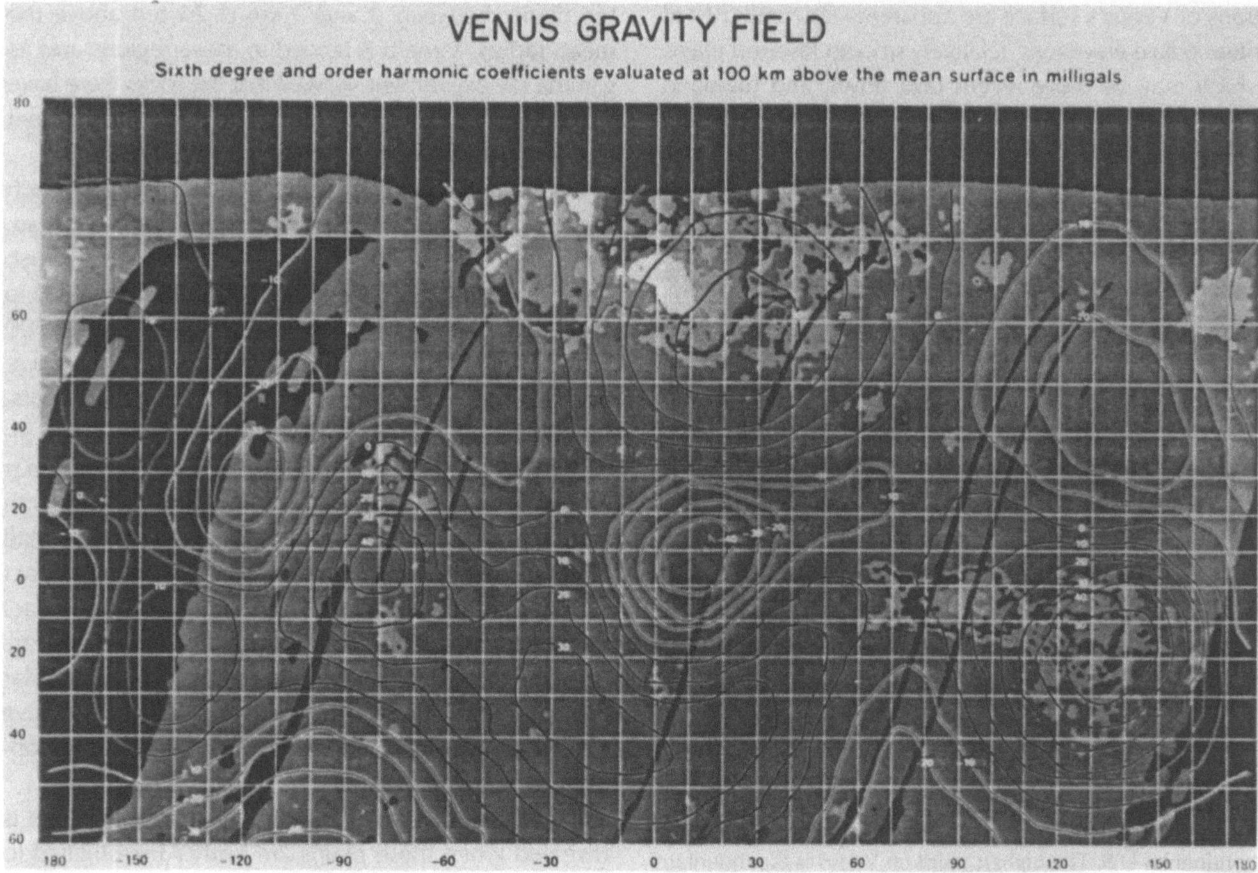

Figure 6.3. There are no big mass concentrations on Venus but the gravity field corresponds with topographical features. Highest gravity corresponds with the Maxwell mountains, and lowest with the lowlands of Venus. (NASA/ARC)

have taken place to reduce topographic effects, and today there is partial isostasy or general equilibrium of crustal masses.

Orbiter's lifting of the veils of Venus (figure 6.4) has revealed a world of great mountains, extensive plateaus, enormous rift valleys, and shallow basins, many of which had been deduced from Earth-based radar images. The wide range of the Pioneer data about the surface confirmed the existence of many features seen by radar from Earth and considerably expanded the coverage of the planet. High-resolution Soviet images further confirmed many of the surface details. However, from the new data many of the earlier radar interpretations had to be revised.

A preliminary description of the probable history of the crust of Venus can be derived from the Pioneer Venus altimetry and images coupled with Earth-based and Soviet spacecraft radar data. Three quite different regions of Venus's surface are apparent—ancient crust at intermediate elevations, relatively smooth lowland plains which may be more recent lava flows, and highland areas which may be continents. Sixty percent of the surface of Venus appears relatively flat in the radar maps; these areas have been termed rolling plains. On them the elevation varies by about 1000 meters (3000 ft) between high and low points. The mean height of this planet-encircling plain is 6051 km (3760 mi) from the center of the planet and has been adapted as a reference as sea level is a reference on Earth.

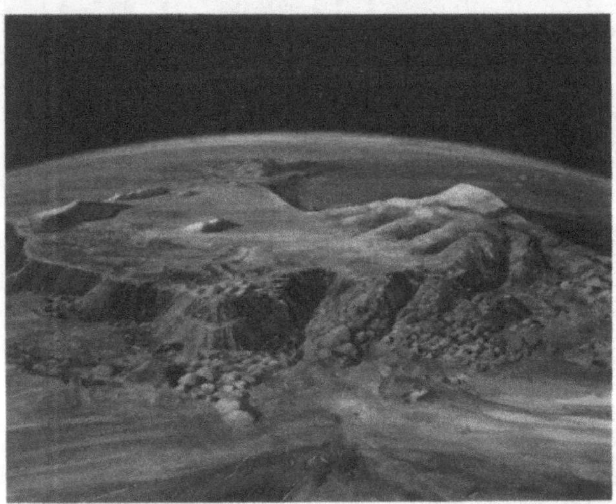

Figure 6.4. Ishtar Terra is the highest and most dramatic continent-sized highland region of Venus. This artist's conception is based on topography measurements made by the Pioneer Venus Orbiter. The continent is about the size of the continental U.S. The highest point on Venus is the mountain massif at the right, Maxwell Montes. (NASA/ARC)

About 16 percent of the surface lies below this mean level, compared with 66 percent of Earth's surface—the ocean basins. Most of the remaining 24 percent of the surface of Venus is only a few thousand feet higher than the rolling plains. Only 8 percent of the planet can be regarded as true highlands, ranging in height to maximum altitudes of 10.6 km (6.6 mi) on one of the highland areas. These highlands may resemble continents of Earth and be "floating" higher than other areas.

Most of the ancient crust of the planet appears to be preserved in the upland plains of Venus, those parts of the planet between 0 and 2 km (1.24 mi) above the mean radius. Venera 8 landed in these regions and its gamma ray experiment showed that the rocks there have a uranium, thorium, and radioactive potassium content that is consistent with a granitic composition.

On the upland plains are circular dark features which may possibly be remains of large impact craters. If so, these plains are the remaining parts of ancient crust. The circular features are about 500 to 800 km (300 to 500 mi) in diameter but very shallow—only 200 to 700 m (650 to 2300 ft) deep. Their shallowness may have been caused by erosion or by lava flooding their floors. Bright spots in the radar images of the craters may indicate that they have central peaks. Earth-based radar mapping reveals other smaller circular features with narrow rims and dark, deeper floors. There are also small circular features which look much like young impact craters from which the material ejected by the impact has produced a surrounding rough area which appears bright on the radar images. These differ from circular features on several highland areas which appear to have extensive lava flows associated with them and are obviously volcanic in origin.

If a full population of craters down to smaller sizes is revealed when these plains are further investigated to higher resolution by a Venus Orbiter Mapper spacecraft, the next American mission to Venus, then Venus may be shown to have preserved ancient high-standing crustal material. Counts of the numbers of craterlike features now revealed produce a crater density curve that aligns with those derived from counts of craters on other terrestrial planets. This supports the viewpoint that a heavily cratered ancient crust is preserved on Venus as it is on the Moon, Mercury, and Mars, but the statistical sample is not sufficient to confirm this.

The lowlands of Venus cover about 20 percent of the surface compared with the terrestrial lowlands, which cover 75 percent of Earth. They also differ markedly from the lowlands of Earth which are the floors of the terrestrial ocean basins. Plateaus and mountains on Venus are as high or higher than those of the Earth but the lowlands are only one fifth the greatest depth of Earth's lowlands. To the radar they appear dark, so they must be smooth or else consist of radar-absorbing material.

An extensive lowland basin, Atalanta Planitia, centered at 170° longitude and 65° N latitude, is about the size of the Earth's North Atlantic Ocean basin. (Except for some features with names that have become established from Earth-based observations, the features on Venus are given female names following the tradition of the name of the planet itself, the only planet of the Solar System with a female name.) The smooth surface of Atalanta Planitia, about 2 km (1.24 mi) below the mean elevation, resembles the mare basins of the Moon. Because there are no circular features that could be impact craters on the lowland areas of Atalanta, the surface may be young. The basin forms part of a large belt of irregular unconnected lowlands—possibly lava-flooded areas—encircling the planet. One interpretation of gravity anomalies is that the crust beneath these plains is thin and of lower density than the crust beneath the upland plains, somewhat similar to conditions on the Moon and Mars. There have been suggestions that these low areas are depressions filled with basaltic lavas, like the mare surfaces of the Moon and some of the plains of Mars.

Although it is not as thick as the crust of Mars and the Moon, the crust of Venus is thought to be thicker than that of the Earth, possibly to the point of shutting off crustal movements of plate tectonics. The lower layer of this crust may consist of heavier basaltic rocks around the whole planet, on top of which is a layer of lighter granitic-type continental rock.

Some interpretations suggest that Venus is one very large tectonic plate compared with Earth's six major plates and several minor ones. Because of the formation of new plate material, Earth's plates constantly expand and grind together and sometimes are forced one under the other in subduction zones, which carry the material back into the mantle and complete the convection cycle which brings heat from the interior of the Earth. Plates also become buckled to form major mountain ranges such as the Himalayas.

On Venus only about 16 percent of the surface consists of low-lying basins comparable to Earth's ocean basins. There is no evidence of features similar to the mid-ocean ridge spreading centers, where molten basalt wells up from Earth's interior to form new crust. By evolving much more crustal material than Earth and producing a thick top layer, Venus could have squelched continuance of plate tectonics, which may have occurred in past periods. But plate tectonics as an active molder of Earth's surface has been accepted for such a short time that we do not have a full understanding of how tectonics might operate on a waterless world such as Venus.

From data received from the spacecraft radar, supplemented by observations by Earth-based radar, scientists have identified major highland features on the cloud-shrouded planet. Two huge continent-sized highland areas are prominent. One is about the size of Australia and contains mountains higher than Mount Everest. The other has a somewhat lower terrain and a size about that of Africa. These highland areas of Venus do not appear to have any circular features that could be interpreted as craters. However, this may be because craters are difficult to detect by radar on rough terrain. The existence of these highlands may imply that there is very little water in the crust of Venus, because at the high surface temperatures water-rich crustal rocks would deform more readily and the highland areas would not persist. Although some circular features have been identified on radar images of the rolling plains, there do not appear to be abundant primordial impact craters on Venus like those on Mars, Mercury and the Moon.

The two highland or continental masses on Venus have been named Ishtar Terra and Aphrodite Terra. (Ishtar was the mythological Babylonian and Assyrian goddess of love, and Aphrodite the Greek goddess of love.) The Soviets, however, claim to have discovered a north polar continent in an area beyond the coverage of Pioneer. Ishtar Terra is located between 30° E and 60° W longitude and 60° and 75° N latitude, and Aphrodite Terra, between 80° and 140° E longitude and 5° N and 15° S latitude. The much smaller elevated region, Beta Regio, located between 40° and 50° W longitude and 10° and 40° N latitude, appears to be a volcanic area associated with a major rift valley system. Beta Regio is probably the youngest region, and its mountains are possibly still forming. Next in age is Ishtar Terra, and the oldest region is Aphrodite Terra.

Points on Ishtar rise to about 11 km (7 mi) and on Aphrodite to about 5 km (3 mi) above the mean radius of the planet. But only 5 or 6 percent of the surface in these "continental" regions is more than 1600 meters (5200 ft) above the mean level; compared with 30 percent on terrestrial continents. The mass of these regions is about 80 percent compensated. Three possible causes are mantle convection placing silicic rocks beneath the highland masses, mantle plumes of upwelling magma producing local differentiation to balance the thickness of the crust, or plate tectonic processes causing continental growth. Continental growth by tectonics does not have supporting evidence of deep subduction troughs or mid-basin ridges which are characteristic of terrestrial plate tectonics. However, the presence of some complex forms of troughs and ridges in many areas may indicate that there have been large-scale motions of the crust.

The highest and most dramatic continent-sized highland region, Ishtar Terra, is a high plateau carrying several mountain ranges. Ishtar Terra is about the size of Australia or the continental United States (figure 6.4). It has the highest peaks on Venus and consists of three geographic units, Maxwell Montes, Lakshmi Planum (named after a Hindu goddess) with mountain ranges of Akna Montes (named after the Mayan goddess of birth), and Freyja Montes (after a Norse goddess) on its northern and northwestern margins, and an extension of the Lakshmi Planum. Lakshmi Planum forms the western part of Ishtar Terra and appears to be a smooth plateau about 3300 m (10,000 ft) above the mean level. Lakshmi is about the same general elevation as the terrestrial Tibetan plateau is above Earth's mean sea level. But it has twice the area of the terrestrial plateau. A bright scarp on the southern boundary may consist of talus slopes of eroded debris along a fault zone. Such a rough surface could account for the strong radar reflection. Lakshmi is bounded on the west and north by mountains ranging upward from 2300 to 3300 m (7000 to 10,000 ft) above the plateau and 5700 to 7000 m (17,000 to 20,000 ft) above the mean level. The western mountains are named Akna Montes and the northern mountains are Freyja Montes. The central area of the plateau is smooth on the radar images and may be covered with relatively young lavas. The huge escarpments around the edges of the plateau drop precipitively to the rolling plains.

On the eastern side of Ishtar the huge Maxwell Montes thrust their peaks high into the Venusian sky (figure 6.5). But unlike terrestrial or Martian mountains they can never be shrouded in clouds because the lowest cloud layer is always many kilometers above them. The high-

Figure 6.5. Seen from the rolling plains, Maxwell Montes might look something like this. The mountain is about 10.8 km (6.7 mi) above the mean level of Venus. It is believed to be an ancient volcanic pile. (NASA/ARC)

est point measured on Venus is in this mountain massif of Maxwell Montes which was discovered by Earth-based radar. This huge area of uplifted terrain occupies the entire east end of Ishtar Terra. Its highest point is 11,800 m (35,300 ft) above the mean level. The highest parts of the massif run northwest to southeast with lower projections extending east and west. On it is a great circular feature which may be a volcanic caldera about 100 km (62 mi) across and 1 km (0.62 mi) deep which is offset on the east flank of the mountain some 2 km (1.2 mi) below the summit. No bright flows radiate from this caldera. The implication is that erosion has smoothed any lava flows. If so, the volcano must be much older than those elsewhere on Venus.

Observations from Earth-based radar and spacecraft radar suggest that the mountain region is very rough, a jumbled terrain which is very different from the smooth plateau to the west. Its radar brightness suggests that the steep slopes of the mountains are covered with rocks. Much of the slopes of Maxwell are, however, bright in the radar images, indicating that they are covered with rocks that scatter the radar, probably because the slopes of the mountains are covered with debris. Polarization data indicate that these slopes are rougher than the very rough floor of the fresh lunar impact crater Tycho, which is the roughest area of the Moon.

If Ishtar consists of basaltic lava flows a large gravity anomaly would be suspected. But the data from Orbiter show a relatively mild positive anomaly. This suggests that Lakshmi Planum consists of thin lavas overlying an uplifted segment of ancient crust, similar to the Tharsis region of Mars. Again this seems to confirm that Ishtar is an ancient volcanic area.

East of Maxwell and extending for 100 degrees of longitude there is a complex topography of ridges and troughs including many closed basins.

Aphrodite Terra (figure 6.6) consists of two mountainous areas separated by a somewhat lower region. Situated almost on the planet's equator, Aphrodite Terra runs almost directly east and west for 9600 km (6000 mi). Unlike Ishtar Terra, the Aphrodite highland region rises to various heights above the mean surface level. The western mountainous area rises 8000 meters (26,400 ft) above the surrounding terrain. The eastern mountains rise 3,300 meters (10,000 ft) above surrounding terrain. As on Ishtar the mountains appear to be very rough. Because Aphrodite does not appear to contain uplifted plateaus or volcanic mountains, it may be older and more degraded than Ishtar Terra.

Figure 6.6. The other large continent-sized highland mass is Aphrodite Terra. This is close to the equator and may be an active volcanic region with fresh young volcanoes. (NASA/ARC)

Scorpion-shaped Aphrodite Terra is about the size of Africa. It has two mountainous areas; on the east, mountains rise 5.7 km (3.5 mi) above the mean radius of Venus. On the west, claw-shaped mountains are about 4 km (2.5 mi) high. Between them are rolling uplands with a topographically complex mountain rising about 3 km (1.9 mi) above it. The mountains have very rough surfaces like those of the Ishtar continent. South of Aphrodite is a large arcuate feature (figure 6.7) called Artemis Chasma. A big circular feature may be an ancient, degraded impact basin.

The bright radar area of Beta Regio is also an interesting region dominated by a large shield volcano, possibly two, and a large trough (figure 6.8). The trough is part of a fault zone that may extend far to the south where two other smaller highland areas (Phoebe Regio and Themis Regio) are aligned. Other highlands, including Asterio Regio, located west of Beta Regio, have a north-south trend. Lava flows extend radially from the volcanic centers, and two Soviet spacecraft landed

Figure 6.7. An arcuate feature southeast of Aphrodite Terra may be an active volcanic feature resembling island arcs of Earth. (NASA/ARC)

directly east of Beta and found basalts there. The largest mountainous features on Beta Regio are Theia Mons and Rhea Mons, both of which are 4 km (2.4 mi) high and appear to be shield volcanoes. (Theia and Rhea are two of the six female Titans, daughters of Gaea in Greek mythology.) A large southward-trending ridge has elevations up to 2 km (1.2 mi). West of Beta Regio is flat terrain with a linear tectonic feature extending 4500 km (2800 mi) to the south-southwest.

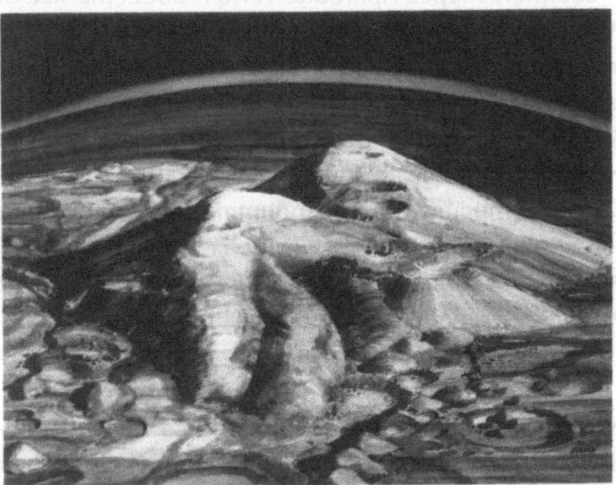

Figure 6.8. Rising from the great plains of Venus about 30° north of the equator is Beta Regio, which is thought to be two huge shield volcanoes, larger in volume than the Hawaii-Midway chain. The two volcanoes have been named Rhea Mons and Theia Mons. Both are about 4 km (2.5 mi) high. (NASA/ARC)

The new information about this region has proven to be of great geological interest. At first from Earth-based radar data Beta seemed to be a shield volcano with a central caldera, but additional information from the Pioneer Venus indicates that it is, in fact, an upland area of young volcanics split by a great rift valley with high shoulders whose nearest terrestrial analog is the Great African rift valley system. Bright radial streaks radiating from these shield volcanoes are suggestive of lava flows which lead to the conclusion that this is a comparatively young geologic feature.

High-resolution radar images of Beta Regio obtained from Arecibo in 1983 showed a large volcano from which lava poured over the surrounds and into a large rift valley. Details of the faults in the valleys were shown clearly by the radar and the 200-km (125-mi) wide valley is identified as one of the largest such features known in the Solar System. It is believed that the valley is about 2400 m (8000 ft) deep at its deepest section. Another large volcano straddles the canyon.

Centered at about 30° N latitude, Beta Regio thus consists apparently of at least two large shield volcanoes situated on a fault line running approximately north–south. This long fault zone covering 10 degrees of latitude connects several other highland features which may be volcanic and are located south of Beta Regio. The two huge shield mountains with smooth radar surfaces cover a north–south distance of about 2100 km (1300 mi).

Alpha Regio (figure 6.9), a plateau within the rolling plains of Venus, is located at 25° S latitude and 355° longitude. It is one of the brightest features on Venus and is elevated about 0.5 km (0.3 mi) above the mean level with a 2-km (1.2 mi) high rim. Its surface is cut by many fractures. Alpha Regio is a rough region with an elevation of about 1800 meters (6000 ft) above the mean level. It is about 25 degrees south of the equator, and radar imaging reveals a rough terrain with parallel fractures. It may combine old and new geologic forms resembling the basin and range structure of the western United States.

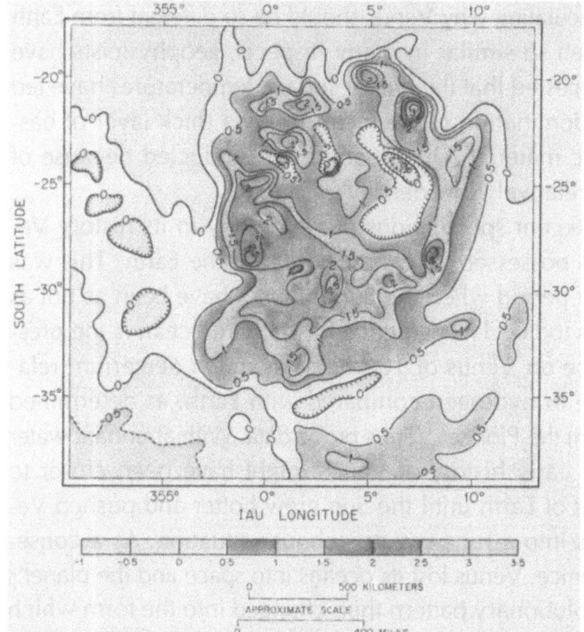

Figure 6.9. A smaller highland group is the roughly circular upland region called Alpha Regio. It is about 1290 km (800 mi) in diameter. This contour map of the area was produced from Pioneer Orbiter data and shows an 800-km (500-mi) wide central depression. Alpha appears as a very bright region on the radar images obtained from Earth. Its distinctness has been used to establish the central meridian of Venus maps at the center of a small oval feature which has been named Eve. (NASA/ARC)

As mentioned earlier, the largest low area on Venus is centered west of Ishtar Terra at 70° S latitude. At its deepest point the great basin is about 3000 meters (10,000 ft) below the mean level. The area is smooth and it lacks any large circular features that might be interpreted as craters. Like Earth's ocean basins it might be young geologically and covered with basaltic lava flows. Alternatively the floor may be covered with windblown debris.

The lowest point on Venus appears to be in a rift valley just east of Aphrodite Terra. Its depth is about 2900 meters (9500 ft) below the mean level, although parts too small to be revealed on the radar images may be deeper. This trench is deeper than the Dead Sea rift on Earth but only about one-fifth the depth of the Marianas Trench in the Western Pacific. It is roughly the same depth as the Vallis Marineris, the great canyon on Mars. While Venus does not appear to have plate tectonics, this rift valley and another parallel to it, both in the region east of Aphrodite, seem comparable to tectonic rifts on Earth where spreading centers are developing on continents to split them apart and form ocean basins between them. In general the jumbled region East of Aphrodite Terra is characterized by high ridges and deep valleys similar to a region east of Ishtar. Both might be evidence of tectonics on Venus.

Many rift valleys have now been discovered on Venus (figure 6.2). They appear to be straight, or gently curved, tectonic features some of which are 5000 km (13,000 mi) long. In various regions they form striking patterns and there is a great concentration of them east of Aphrodite and also east of Ishtar. They probably result from regional distortions and may represent the early stages of plate tectonics.

Geophysics points to a somewhat different interpretation of the radar images. The gravity field of Venus as mapped by the Orbiter matches closely the topography, much more than for Earth where, for example, in the Pacific Ocean there are large gravity anomalies that may be the result of dynamic processes. But the anomalies of Venus seem to be closely associated with the topography. Venus is nevertheless more like Earth than like the Moon and Mars. But the mystery is that geologically the planet looks like a planet on which the heat engine has operated differently and the planet has settled down to a fairly changeless existence, as has the Moon and perhaps Mars. On the other hand the gravity distribution of Venus makes it look less like Earth. In terms of stress implication, Venus and Earth are intermediate between the Moon (much less) and Mars (much more).

While the total difference in elevations is about the same on Venus as on Earth, the smooth surface of Venus is not likely to result from the absence of plate tectonics on Venus because much rougher planets—the Moon and Mars—do not have plate tectonics either. It could, however, result from the high surface temperatures of Venus which allow the crustal rocks to deform easily and thereby smooth elevation differences. A large-scale effect of this kind would work against the preservation of large highland areas but not of smaller features.

That tectonics has modified the surface seems clear, because craters formed by the impact of bodies from space seem to have been destroyed or considerably modified. The surface seems to have been disrupted tectonically, and as already mentioned some regions show faults and folded mountains. There are relatively very few features on Venus that can be readily identified as impact craters. The large-diameter circular features, many of which have been identified on Earth-based radar images, are extremely shallow.

East of Ishtar there is evidence pointing to tectonics in a large region, extending from 40° to 14° longitude and from 50° to 75° N latitude, which consists of complex ridges and troughs, probably disrupted by extensive faulting. It appears to be the most tectonically disturbed region of Venus. There have been speculations that this region is possibly one where plate tectonics started or where a plume of hot magma rose through the mantle to produce a thickened low-density crust. A similar area is located east of Aphrodite Terra.

Other features also suggest tectonic activity on Venus: vertical uplift at Lakshmi Planum, and the northern and western mountainous ridges marginal to Ishtar. These ridges on the Ishtar Terra may be due to plate motion, but there is no evidence for integrated plate tectonics on Venus. The development on Venus of thin-crusted lowlands and thick-crusted highlands suggests that Venus experienced a period of widespread mantle convection early in its history. The resolution of the Earth-based radar and the Pioneer Venus images is, however, not sufficient to entirely exclude the existence of plate tectonics on Venus. But it is sufficient to say that if plate tectonics do exist on Venus, they are grossly different in character from the plate tectonics of the Earth.

Venus seems to be quite different from any of the other terrestrial planets. It seems to have signs of regional displacements which may be evidence of incipient, rudimentary, or past plate tectonics. Development of plate tectonics may have been stopped because Venus lacks water, but there is no proof that the presence of much water has anything to do with plate tectonics. Speculating why Venus should be so different from Earth when so similar in many respects, geophysicists have suggested that the higher surface temperatures have led to domination of the tectonics by a thick layer of basaltic material which cannot be subducted because of its relatively low density.

Recent speculations are that early in its history Venus possessed an ocean similar to the Earth. This was at a period when the Sun may not have been as hot as it is today. The evidence for such an ocean is the presence on Venus of 100 times as much deuterium relative to hydrogen compared with Earth, as determined from the Pioneer Venus probe data. With abundant water the early history of Venus might have been closer to that of Earth until the Sun grew hotter and pushed Venus into a runaway greenhouse situation. As a consequence, Venus lost its oceans into space and the planet's evolutionary pattern then changed into the form which led Venus to its present state. With high-resolution radar mapping of Venus by an advanced spacecraft or more widespread optical images of the surface from balloon probes beneath the clouds, evidence of ancient oceans, such as beach lines, might be found.

After water was lost on Venus plate tectonics possibly ceased. The two "continental" masses, Ishtar Terra and Aphrodite Terra, appear to be supported, probably by hot spots beneath them, unless they are geologically very young features and have not had time to deform back into the median plains.

One interpretation of Venus's topography is that the equatorial highlands of Venus—Beta Regio and Aphrodite Terra—are the equivalent of Earth's mid-oceanic ridges from which the ocean crust spreads. Without subduction the crust of Venus would not be able to spread and highland masses would rise along the equatorial zone. Because these highlands exist today it appears that the activity continues as hot-spot dynamics. Ishtar Terra, by contrast, might be a true continent formed as a product of plate tectonics early in the history of Venus. Further support of this interpretation comes from the ridges on the east of Ishtar which look somewhat like the terrestrial island arcs such as the Aleutian chain.

If so, the major tectonic activity on Venus today is at

Beta Regio and Aphrodite Terra. Again there is some support for this viewpoint in that most of the signals assumed generated by lightning discharges originate at Beta Regio and Aphrodite Terra, which would suggest the presence of volcanic plumes. Because there is little overturning of the lower atmosphere of Venus and no precipitation or thundercloud type activity, it seems most likely that the lightning discharges are produced in volcanic clouds over active volcanoes (figure 6.10).

New evidence for active volcanoes on Venus was presented early in 1984 as a result of comparisons between a variety of findings from the Pioneer Venus data. University of Colorado scientists used five years' analysis of atmospheric data from the Pioneer Venus Orbiter to show that concentrations of sulfur dioxide in the atmosphere of Venus increased more than 50 times in 1978 and declined from then on. The abnormally high amounts of sulfur dioxide are attributed to a major volcanic eruption on Venus, which occurred shortly before Pioneer Venus began its explorations in 1978. This is analogous to major increases in atmospheric sulfur dioxide which occur following volcanic eruptions on Earth.

In addition Earth-based radar images of Beta Regio in 1983 show geologic details interpreted as relatively recent volcanic activity because of absence of impact

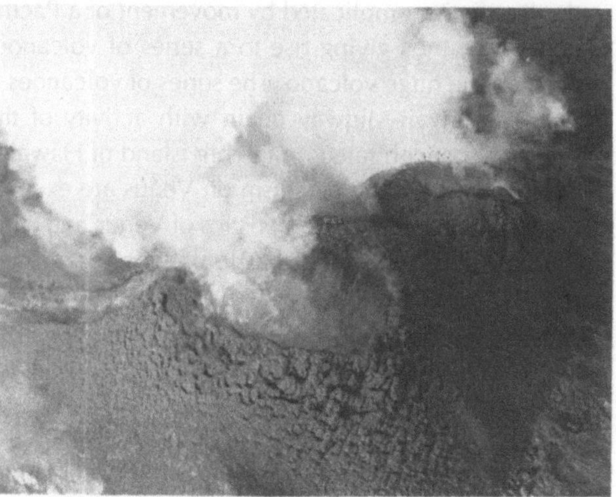

Figure 6.10. Evidence has been mounting that there are currenty active volcanoes on Venus building shield volcanoes like those in Hawaii. (Photo Cindy Kirkpatrick)

craters. Rifting and volcanism appear to be significant processes on Venus from the Earth-based radar images.

Volcanic activity has been suspected on Venus following the discovery of whistler-type radio signals by Russian and American science teams. The lightning responsible for such signals in the terrestrial atmosphere is often generated in volcanic plumes. On Venus the lightning concentrates near the surface of two mountainous regions—Beta Regio and Atla Regio (figure 6.11), both believed to be shield type volcanoes similar to the Tharsis volcanoes of Mars. Radar altimetry images, general topography, and gravity sensing data have convinced some geologists that Beta Regio and Atla Regio are both volcanic mountain regions; they are certainly of the correct size and shape. Large positive gravity anomalies indicate that the volcano is less than 1 million years old. Atla Regio, in the "Scorpion's Tail" region of Aphrodite Terra, and Beta Regio are on the equatorial belt that some geophysicists regard as an equivalent of the terrestrial spreading centers where heat and hot magma rise to the surface from the interior of the planet.

The gravity sensing data indicate that these areas are formed of new rock recently extruded from the interior of Venus. As already discussed, Beta Regio has been interpreted as consisting of two enormous shield volcanoes, almost 2100 km (1300 mi) in length, larger than the entire Hawaii-Midway chain and comparable with Olympus Mons on Mars. This area appears to be the most volcanically active area of Venus, and images from the Venus radar and ground-based radar show radiating bright rays on Beta Regio which are indicative of young lava flows. Also one of the Venera landers found volcanic basalts on the flanks of the mountains.

Beta Regio is believed to be located over a powerful upflowing convective plume which rises from deep in the interior magma of the planet. In fact, Beta Regio might dwarf all other volcanoes in the Solar System in volume, being larger than the huge Martian volcano, Olympus Mons, but not as high perhaps because of Venus's stronger gravity.

Recent Earth-based radar images indicate that Maxwell Montes on Venus's Ishtar Terra is also a large volcano, but one that does not have a strong gravity

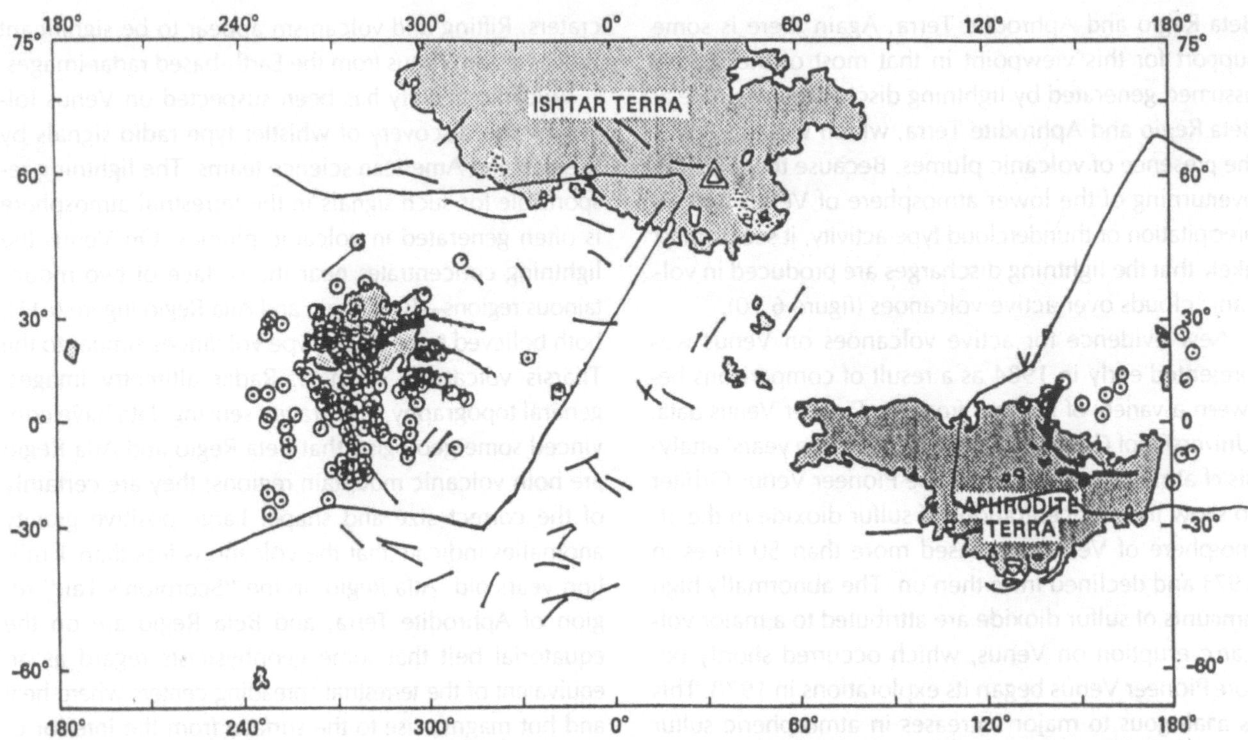

Figure 6.11. Spacecraft have observed radio signals that appear to be originating from two areas of Venus, Beta Regio and the eastern end of Aphrodite Terra. Topographically, also, these are most likely to be volcanic areas. (NASA/ARC)

anomaly, and now seems extinct. It shows no signs of recent activity.

The huge volcanoes on Venus appear to be related to the planet's global heat balance. Although Venus appears to have the same internal heat source as Earth, our planet is able to vent its heat from many points, particularly along the mid-ocean ridges from which new crust spreads. Some scientists believe that Venus, unlike Earth, has no mobile plate tectonics today and hence there are no extensive spreading regions where new crust forms and is forced apart as liquid magma comes to the surface almost continuously. Instead there are a few "hot spots" over upwelling plumes of magma rising from the mantle.

The volcanic outpourings of Beta and Atla Regios appear to be the youngest form of heat escape on Venus. Upwelling plumes in the internal magma produce these hot spots which break through the thick rigid crust and vent most of the planet's internal heat at these two places. The process is similar to that which forms the big volcanoes on Hawaii and formed the Tharsis volcanoes on Mars. However, the terrestrial analog is not exact because the surface manifestation of the plume in the Pacific is complicated by movement of a Pacific plate over it, thus giving rise to a series of volcanoes instead of one huge volcano. The series of volcanoes is today the Hawaii-Midway chain with activity of the plume now concentrated on the big island of Hawaii.

The findings about volcanism on Venus are exciting because the physical characteristics of Venus and Earth are almost identical. Venus has long been regarded as Earth's twin because of the two planets' similarity in size, mass, gravity, and distance from the Sun. But while Venus may well be volcanically very active, this activity differs markedly from that on Earth at the present time.

As has been noted, it seems from the observations that a major volcanic eruption occurred on Venus in 1978, which forced enormous quantities of sulfur dioxide into the atmosphere of Venus together with many small particles that contributed to a general haze. The sulfur dioxide rapidly formed into small aerosol particles of sulfuric acid similar to acid rain on Earth. Increasing haze also formed over polar regions. Pioneer Orbiter although not in orbit during the eruption has been recording the slow decay of the effects of the eruption ever since. The unusual increase in sulfur dioxide was also observed by independent ground observations in 1978. A check back to earlier records shows that a similar increase in sulfur dioxide in the atmosphere of Venus occurred in the 1950s. As the sulfur dioxide decreased in the atmosphere, the polar haze also faded.

The volcanoes could blow sulfur up to as high as 70 km (40 mi) through the extremely dense atmosphere of Venus. Such an eruption would have at least ten times more energy than any eruption on Earth in the last 100 years, much larger than the eruption of Krakatoa in 1883. While some scientists have expressed doubt that Venus is volcanically active today, the Soviet spacecraft Venera 15 and 16 found volcanic cones elsewhere on their radar images, for example, in the Metis Regio west of Ishtar Terra. Moreover, the big radio astronomy telescope at Arecibo, Puerto Rico has also imaged volcanic cones and craters on the planet.

If Venus is volcanically active today, it will rank as the third most active terrestrial-type planet, after Jupiter's satellite Io and Earth. There is currently no evidence that the Martian volcanoes are still active.

The presence of much primordial argon, and the slow retrograde rotation of Venus, suggest that at some time in the past Venus was catastrophically impacted by a fairly large body, which could have been a satellite. If this impacting body had been close to the Sun during the formation of the Solar System Venus ought to have acquired the argon isotope which is abundant on Venus today. One can speculate that, just as Earth acquired the Moon in the early days of the Solar System, Venus also may have acquired a large satellite—but into a retrograde orbit. After such an impact any other satellites moving on prograde orbits would have been drawn in to Venus by tidal friction.

Unfortunately we do not know enough about the early history of Earth to determine whether or not Earth and Venus might have followed parallel evolutionary paths with plate tectonics, subduction, and recycling of water back into the interior during the first several billion years of planetary evolution. That might well have been the case, with Venus and Earth existing for many millions of years as almost identical twins. But loss of water by Venus could have caused significant changes not only in the atmosphere but also in the lithosphere of the planet. With no water to recycle back into the interior the characteristics of the magma would change and the type of tectonics would also change. On Venus the loss of water would enhance high surface temperatures in resisting subduction of crustal material back into the mantle. Plate tectonics would slow and stop. At spreading rates comparable to those on Earth today, the whole of Venus would have been choked with continental crust within a few hundred million years. But although plate tectonics would end, the planet could continue volcanically active to the present time.

Venus is probably a planet like Mars with a single planetwide tectonic plate, and the rolling plains are probably the planetwide crust.

While it might be, and was at one time accepted, that planets of the same mass and dimensions evolve along similar lines, the exploration of Venus has shown that this is not so. There are many as yet unknown factors which determine how planets evolve, and similar planets can evolve quite differently. The question still remains, nevertheless, whether planets that have evolved in one direction can be set into another direction by some catastrophe; particularly, could Earth be diverted into a Venus-like planet? Could the impact of water-rich objects on Venus make the planet more Earth-like? Despite the many expeditions to Venus we still do not have answers to these very important questions. Urgently needed are petrologic and geochemical sampling of the surface of other worlds, particularly Venus and Mars. At present we only have definitive information of this nature about the Earth and its Moon.

Comparisons of the evolution of Earth and Venus are given in table 6.1 on the basis of some modern scenarios. Information about the surface at four locations on Venus has become available from the Soviet landers which photographed and sampled the surface. Veneras

Table 6.1. Evolution of Venus and Earth Compared

	Venus	*Earth*
1) Accretion	Major impact stops rotation	Random impacts
2) Heating	Core forms	Core forms
	Outgassing of volatiles	Outgassing of volatiles
	Atmosphere formed	Atmosphere formed
3) Secondary Bombardment	Cratering of crust	Cratering of crust
4) Some crustal heating	Degradation of some craters	Degradation of some craters
5) Crust Stabilizes	Tectonics spreading subduction water loss begins	Tectonics spreading subduction no water loss
6) Plate tectonics	Subduction ends Surface chokes Single plate formed Hot-spot tectonism continues	Subduction and multiplate spreading center tectonics continues

9 and 10 gave the first views of the rockstrewn desert of Earth's errant twin. Venera 9 landed on sloping terrain to the east of Rhea Mons at the border between a mountainous region of the shield volcano and the rolling plains surrounding it. The rocks are rather large at the site, their diameters ranging from 50 to 70 cm (20 to 30 in). They were slabs of rock rising up to 20 cm (8 in) above the soil and they seem to be products of the breakup of larger rocks with only a small amount of subsequent weathering. There are also a few rounded rocks. Between the rocks is a much darker soillike surface with small pebbles embedded in it.

By contrast, the view from Venera 10, which landed on the flatter ground of the rolling plains southeast of Rhea Mons, shows stratified layers of tilted bedrock between outcrops of which are pockets of dark soil. The light-colored rocks appear quite flat with relatively smooth surfaces and rounded edges. They appear to have been exposed to considerable weathering, possibly by wind-driven soil and pebbles. There is also a suggestion of honeycomb weathering, which may be indicative of chemical disintegration of the rocks. A few small rocks are seen embedded in the dark soil.

Russian scientists state that the rocks are basaltic, which fits into the view that Rhea Mons is a shield volcano. But the surface does not look like fresh lava flow. One possibility is that weathering occurs very rapidly on Venus so that fresh lava does not remain for long on the surface before it is acted upon.

Significantly, the views on the surface of Venus are not so very different from those on the surface of Mars at the Viking lander sites, nor for that matter from views which can be obtained in many arid or desert regions of Earth. The Veneras 8, 9, and 10 carried instruments to analyze some of the surface materials and found that the material from the Venera 8 site north of Hathor Mons has the largest concentration of natural radionuclides. This seems analogous to Earth's ancient crustal materials. Analysis of material at the Venera 9 and 10 sites, however, indicates that the rock there is younger than at the Venera 8 site, again pointing to the Beta Regio as being a young volcanic area. Density measurements of the rock also point to basaltic composition at the Venera 9 and 10 sites. This corresponds somewhat to Earth's ancient crust of continental rocks which have a higher radionuclide content than the rocks of the younger oceanic crust.

Veneras 13 and 14 landed at locations on Venus which are more like the Venera 10 site than the Venera 9 site. Venera 13 landed between Phoebe Regio and Nava Planitia, and Venera 14 landed a short distance to the east in Nava Planitia. Again the flat rocks are prominent feature of the landscape in the images returned to Earth. Again there is the dark soil between the rocks. The Venera 14 site is characterized by flat rocks covering most of the visible surface and with cracks and a few small pockets of soil. The Venera 13 site has more areas of soil visible. At neither site do the rocks project very much from the general level of the surface. In looking at the pictures one is reminded very strongly of pictures of the bottom of terrestrial oceans. Indeed, the pressures are similar on the surface of Venus and in the shallower terrestrial oceans. Motions of the atmosphere of Venus close to the surface may behave more like ocean currents than terrestrial winds in moving and depositing sediments and eroding exposed rocks.

These spacecraft carried drills to sample the rocks. The Venera 13 sample was of a basaltic rock with a high fraction of potassium. The Venera 14 sample had much less potassium.

While the views of Venus show a consolidated regolith there is no data as to how thick this regolith may be. Radar probing of Venus suggests that the regolith is not as thick as that of the Moon. In fact, it is probably quite thin and consolidated. The low-porosity, consolidated rock suggests surface materials very similar to the Earth rather than the Moon, from which it might be assumed that the surface of Venus has evolved along lines similar to the Earth but modified by the lack of water on Venus and the difficulty of subducting new crust.

7

MEANINGS AND MYSTERIES

If we understand the atmosphere and surface of Venus we will have a better chance of understanding the evolution of Venus and of the other planets of the Solar System. The atmosphere, for example, contains what may be important clues about the evolution of the planet, because it contains material that has outgassed from the interior. As we have seen, the small quantities of isotopes of rare gases are often more important indicators of evolutionary processes than the major constituents of the atmosphere in unraveling the tangle of conflicting speculations.

There were three broad and basic theories to account for how planets obtained their atmospheres; capture after formation, capture over an extended period, and accretion of many small solid bodies that have been termed planetesimals.

The first theory assumes that the planets formed from a solar nebula and obtained their atmospheres by progressive condensation from that nebula. However, the ratios of isotopes of rare gases to each other and to carbon and nitrogen are far from what would be expected if that was the way the planets formed.

The second theory assumes that the planets originated from materials that were short of the volatiles needed to form an atmosphere, but later acquired these volatiles from other parts of the Solar System as bodies rich in volatiles were gravitationally perturbed to enter into the inner Solar System. This theory does not account for the mysterious increase in primordial argon as one moves inward through the Solar System; for example, Venus has many times more primordial argon than Earth, which has many times more than Mars.

The third theory assumes that during the formation of the Solar System many smaller bodies, termed planetesimals, were formed from grains of matter. These contained all the volatiles needed to provide the planets with atmospheres. In this scenario the planetesimals collided with each other and accreted into planets. Later, as the planets heated, the volatiles were released from their interiors to form atmospheres. This theory requires that temperatures not vary greatly with increasing distance from the Sun.

A major question is why the atmospheres of Venus, Earth, and Mars, which chemically seem to be somewhat similar planets, are quite different. Venus has an atmosphere that is about 100 times as dense at Earth's while Mars has an atmosphere about one hundredth the density of Earth's. The major constituent of the atmospheres of Venus and Mars is carbon dioxide, while that of Earth's atmosphere is nitrogen. The reason for the great differences appears to be the chemical reactions which have occurred during the evolution of each planet and the relatively slight differences in distance from the Sun. Venus was just sufficiently closer and hence warmer to pass into a runaway greenhouse situation. Mars was just sufficiently distant and colder to develop into a refrigerated world. In the runaway greenhouse of Venus liquid water was lost by evaporation into the atmosphere and subsequent dissociation. In the refrigerated world of Mars, liquid water was converted into ice and buried in the regolith to form a deep permafrost. On Mars even much of the carbon dioxide was frozen into polar caps. On Earth liquid water remained as oceans to allow the development of life and subsequent biological processes which changed the primitive atmosphere. In addition, the presence of liquid water on Earth permitted chemical reactions to trap much of the primitive carbon dioxide in rocks. Even on Earth, however, there have been times when large quantities of water became trapped in extended polar caps and glaciation of continents with consequent reduction of the mean sea level.

The presence of liquid water has allowed Earth's rocks to be continually weathered, thereby exposing new surfaces to the atmosphere and maintaining chemical activity between the surface and the atmosphere. Water may also have played a role in permitting plate tectonics and volcanic processes which have cycled crustal materials and brought new rocks to the surface for weathering. The radar images of Venus show evidence of continental masses and volcanic areas, possibly of tectonic activity too. Mars also has volcanoes but shows very little evidence of movement of crustal plates on a planetwide scale as on Earth. The only current weathering processes on Mars and Venus, however, seem to be those resulting from winds, although there is evidence of water erosion having occurred on Mars in the past. Our details of the surface of Venus are not sufficiently resolved to be able to identify any evidence of water activity having occurred there.

One way in which planetary atmospheres can be changed is by the escape of molecules into space from the top of the atmosphere. Controlling factors are the gravity of the planet and the speed of the molecules or atoms of the atmospheric gases. The former depends upon the mass of the planet and the latter upon the temperature. The velocity at which a molecule can escape from a planet is called the escape velocity. It is the same velocity that a spacecraft has to achieve to escape into space. For Earth this velocity is about 11.2 km/sec (6.95 mi/sec). It is slightly less for Venus and only about 5.0 km/sec (3.1 mi/sec) for Mars. The velocity which molecules or atoms can attain depends upon each molecule's temperature and mass. Less massive atoms and molecules reach higher speeds for a given temperature than heavier atoms and molecules. While water molecules may not easily attain escape velocity from a planet the size of Earth or Venus, a water molecule located high in a planet's atmosphere can be dissociated by solar ultraviolet radiation into oxygen and hydrogen. The oxygen may not be able to escape, but hydrogen can. In this way a planet can easily lose much of its water.

Fortunately for living things, Earth has a cold trap in the stratosphere which stops water vapor from rising high enough on the terrestrial atmosphere for molecules to be broken down into hydrogen and oxygen by solar ultraviolet radiation. The solar radiation is stopped by an ozone layer above the cold trap. Consequently Earth has been able to keep its oceans. Mars has kept its water in a deep freeze. But Venus was just that much warmer for water vapor to rise high in its atmosphere and be dissociated so that the hydrogen could be lost into space.

With much water vapor in the early atmosphere of warm Venus the greenhouse effect was reinforced. The water vapor trapped solar radiation and increased atmospheric temperature further to evaporate more ocean water at an accelerated rate. In this way the planet would quickly move into a runaway greenhouse and its oceans would be lost forever. The problem, as mentioned earlier, is what happened to the oxygen. Perhaps it, too, escaped. Or maybe some escaped while the rest oxidized surface rocks.

It is instructive to compare the view of the terrestrial planets before space exploration with that today. In the early 1960s Venus and Mars were seen as much like the Earth. On Mars the white polar caps displayed seasonal changes and they were interpreted as caps of water ice. At other latitudes on Mars seasonal changes were observed in the dark areas, which were interpreted as being due to vegetation. The length of the day on Mars and the tilt of the planet's axis were similar to those of Earth, so that Mars experienced seasons, thereby enhancing the terrestrial analogy and the erroneous idea of there being Martians of an older civilization than any on our planet.

Much less information was available concerning Venus because its surface was hidden from observation. Scientists knew that the atmosphere was rich in carbon dioxide and that the planet emitted a substantial flux of radio waves which could indicate that the surface was at a high temperature. But that was about the total of our knowledge. As a consequence speculations about Venus were even wilder than those about Mars. Some scientists stated that Venus was a moist swampy world, teeming with life, or a warm world enveloped by an ocean of carbonic acid. Others speculated that Venus was cool and Earthlike with surface water and a dense ionosphere that was responsible for the radio emissions.

According to yet another imaginative view, Venus was a warm planet with a massive atmosphere in which intense thunderstorms and fantastic lightning displays flashed through violent rainstorms. An entirely opposite viewpoint speculated that Venus had cold polar regions with 10-km (6.2-mi) thick ice caps and a hot equatorial region. More bizarre were viewpoints of Venus as an extremely hot, cloudy planet with puddles of molten lead and zinc at the equator, and seas of bromine, butyric acid, and phenol at the poles. Closest to what is known of Venus today was a speculation that the planet might be a hot, dusty, dry, and windy global desert.

With a number of successful space missions to Mars accomplished, and with them the ability to examine the surface minutely through a relatively clear atmosphere, planetologists were able to define its characteristics more precisely. Gone were all speculations about areas of vegetation and a globe-encircling system of "canals," and in their place the surface was recognized as widely variegated with ancient crustal material and more recent lava flows, with vast canyons and mighty volcanoes, with ancient craters and terrain that seems to have been molded by deluges. The seasonal ice caps were recognized as caps of frozen carbon dioxide with much smaller permanent caps of layered dust and water ice surrounded by regions of dune fields. Extensive regions of chaotically tumbled blocks suggested areas where vast quantities of permafrost had melted and released deluges of water. Great faults implied the beginning of plate tectonics, and in debris-litered basins vast deserts of sand dunes were seen to stretch for hundreds of miles.

Mars is a cloudy planet, but its terrestrial types of clouds offer many clearings through which surface details can be observed. It is also a dusty planet which is subjected to global sand or dust storms. Layered terrain observed at the poles and on the sides of deep canyons suggests that geological periods have occurred when climatic conditions were much different from those on Mars today.

In all, the new Mars which has resulted from exploration by spacecraft has evolved in several quantum jumps which brought great changes to our conception of the planet. We have seen that a planet has to be studied exhaustively and completely to gain a true understanding of it. Many times in the past scientists jumped unwisely to conclusions about Mars which were proved to be quite false by subsequent missions. Something as complex as a planet cannot be interpreted on the basis of a few snapshots from one or two spacecraft missions. For example the Mariner 4 pictures missed the rift valleys, which led to the conclusion that Mars was Moon-like—a cratered dead world; the subsequent flybys of Mariners 6 and 7 completely missed the volcanic areas of Mars, and reinforced the earlier incorrect conclusion. Even as late as 1970 scientists were stating that Mars was a much less active world than it is now known to be. Had the exploration of Mars ended with Mariners 6 and 7, which imaged less than half the planet at relatively low resolution, our textbooks and science would have been very misleading today.

The same is true of Venus. Our explorations with three Mariners and with Pioneer Venus have cleared up many of the mysteries about the planet, and the Soviet Venera missions have considerably expanded our knowledge, but there is still much to be learned. To be on the surface of Venus would almost be like being beneath a hot dry ocean, so thick is the atmosphere (figure 7.1). Circulation is undoubtedly very slow near the surface and is more like ocean currents than the surface winds of Earth. Major important differences among Earth, Venus, and Mars are connected with the nature and thickness of their atmospheres. A great difference between Earth and Venus is that the atmosphere of Venus is nearly 100 times as massive as that of Earth and consists mainly of carbon dioxide. The other great difference is that Venus has negligible water compared with Earth. As a direct result of this Venus does not have life on its surface. The oxygen which plant life produced from the terrestrial water enabled animal life to gain a foothold on our planet to complete biological cycles and thereby set the stage for large-scale conversion of carbon dioxide to oxygen and permit life as we now know it to flourish. There are two other major differances between Earth and Venus, both of which are still unaccounted for. Unlike Earth's familiar 24-hour day, Venus rotates on its axis once in 243 days, in a retrograde direction. Earth's axis of rotation is tilted with respect to the poles of its orbit, the axis of Venus is very close to vertical.

Figure 7.1. The surface of Venus, especially its lowlands, may be more like the floors of the ocean basin of Earth, in which the atmosphere moves sluggishly like ocean currents. (Photo NASA/ARC)

A major mystery about Venus is why the planet should rotate so slowly and in a direction different from the other terrestrial planets. One possibility is that Venus encountered another astronomical body either in a close approach or an actual impact. One theory suggests that Venus may have had a satellite moving in a retrograde orbit which decayed and struck the planet, thus slowing its rotation. Or perhaps a body from outside the Solar System, or one of the big planetesimals from the outer Solar System, hurtled through the inner Solar System and gravitationally altered the rotation period of Venus, and its direction.

Another mystery is the closeness to tidal coupling with the Earth by which Venus always turns the same hemisphere toward Earth at inferior conjunction when closest to Earth. Almost, but not quite! The tidal coupling between Earth and Venus is very small and seems hardly enough to lock the two planets together and maintain the lock over eons of time.

As mentioned earlier Mars's period of rotation and axial tilt are very similar to those of Earth. But Mars is smaller and its atmosphere is only 1/100th that of Earth. By contrast with Venus there seems to be much water on Mars, but not in the form of oceans, although this may not be true of past ages on the planet. Today the water appears to be locked in polar caps and a permafrost that may be miles thick.

Whether or not Venus ever possessed abundant quantities of water is still in question. At one time the amount of the heavy isotope of hydrogen measured in the atmosphere of Venus was taken to mean that the planet once possessed much water which had somehow been lost. Most of a planetary ocean of water might be lost in several ways:

- By photodissociation water molecules can be split into hydrogen and oxygen; the hydrogen reaches escape velocity and leaks into space while the oxygen combines with surface materials.
- Redox reactions can occur in which water molecules react with molecules of carbon monoxide to produce carbon dioxide and hydrogen which again leaks into space. (The problem with this theory is that there would have to be initially the correct proportions of carbon monoxide and water to remove all the water, and this is regarded as being most unlikely.)

- Reactions can occur with surface materials, in which water can combine with iron oxide and release hydrogen to be lost into space.
- If very active plate tectonics occurred, water might be recycled back into the interior of the planet where high temperatures would dissociate it and allow the hydrogen to leak off the planet.

While much water can be lost by any one of these processes, it is difficult to explain by any one mechanism how all the water could be lost. It may be that several of the processes have occurred on Venus.

Oxygen is also important to the question of what happened to the water. If water molecules were broken down into hydrogen and oxygen, the disappearance of the oxygen has to be explained, since very little of this gas is present in the atmosphere of Venus today. No completely satisfactory explanation is yet available for what happened to the oxygen.

Another major difference among the terrestrial planets is the present condition of their surfaces. Studies of Earth were revolutionized by the discovery that the terrestrial crust consists of huge plates that are propelled by basaltic flows originating at mid-oceanic ridges. To understand the basic mechanisms responsible for this mobility of the crustal plates of Earth, it is useful to study the other terrestrial planets on which such movements are taking place, or have been arrested or never developed. A group of large Martian volcanoes lie on an arc, as many terrestrial volcano chains also do. Mars also has, like Earth, large basins filled with flows of basaltic lava. Separating these basins are higher continental masses which are split by rifts. There are many features suggesting a parallelism between Earth and Mars. Radar surveys of Venus, from orbit and from Earth, have shown similar parallels between Venus and Earth: basins with presumably basaltic lava flows, volcanoes, high continents, and great rift valleys.

Even the Moon possesses raised continental areas, basaltic flows covering great basins, and volcanoes, and the same appears to apply to Mercury from observations by a single spacecraft, Mariner 10, after it had flown by Venus.

The upper atmospheres of those terrestrial planets possessing atmospheres also differ. Of great practical importance is an understanding of the upper atmosphere of our planet. The interrelationship of a planet's atmosphere with interplanetary space and the solar wind is important to understanding the dynamics of the atmosphere and how a planet receives energy from the Sun to drive its weather systems. An especially important region in the terrestrial atmosphere is that known as the thermosphere. This is the region where the transition from air to space begins and where incoming solar radiation first begins to exchange energy with the planet's atmosphere. The thermosphere acts like a giant screen protecting the lower atmosphere and surface from incoming charged particles and small solid objects as well as certain short wavelengths of solar radiation that are inimical to biological systems. The interception of these invaders from interplanetary space is manifested as auroral displays and meteors, popularly referred to as northern and southern lights and shooting stars respectively. The interaction of ultra-short-wavelength-radiation produces ionization in the high atmosphere.

The study of Earth's upper atmosphere has become very important during the last few decades, although exploration of the region began as long ago as 1892 during the First International Polar Year. After the discovery early in this century of ionized layers in the upper atmosphere that could return radio waves back to Earth, most of the exploration was by use of radio waves of various frequencies to sound the upper atmosphere. In the 1940s scientists started to send instrument-carrying rockets to high altitudes. The objective was to sample directly the gases and ionized particles of the upper atmosphere and make measurements of temperature and pressure there. More recently observations were made from above the appreciable atmosphere by satellites and into the thermosphere by powered satellites traveling in elliptical orbits that allowed the spacecraft to dip into the atmosphere at perigee and still maintain orbit.

As discussed in earlier chapters, scientists identified a number of distinct regions in Earth's atmosphere—the troposphere, stratosphere, thermosphere, ionosphere, mesosphere, and exosphere. Beyond the atmosphere is a magnetosphere. As also mentioned earlier, the divisions of the atmosphere into troposphere, stratosphere, and thermosphere are based on the thermal properties of these regions. For example, the troposphere is the region of weather, where heating comes from relatively long wavelength solar radiation being absorbed by the Earth's surface, and heat passes to and from the atmosphere in large quantities as water enters and leaves the atmosphere by evaporation and subsequent precipitation. The stratosphere is a region of relative calm where there is little if any turbulent mixing. Within the stratosphere is an ozonosphere where a layer of ozone absorbs incoming solar ultraviolet radiation.

The mesosphere extends from the top of the stratosphere to a height of about 80 km (50 mi), with a low temperature layer at its top. Higher still the temperature increases because of absorption of short ultraviolet radiation from the Sun as contrasted with longer wavelengths which are absorbed in the upper stratosphere. In the thermosphere short-wavelength solar ultraviolet radiation is absorbed, so the top of this region has a high temperature. The ionosphere is created by this high-altitude absorption.

Three well-defined layers of ionization were recognized from radio sounding of the ionosphere. These layers change in the amount of ionization and in their heights diurnally and seasonally. They also change with variations in the stream of energetic particles coming from the Sun to the Earth. There is close coupling among the ionosphere, the thermosphere, and the magnetosphere. The terrestrial ionosphere is compared with that of Venus in figure 7.2.

The upper atmosphere has often been referred to as the frontier to interplanetary space. It also might be regarded as the protective membrane of the living cell of our Earth. It allows the passage of life-supporting energy from the Sun while screening out the harmful radiation, and it retains within the cell the necessary environment for living things to flourish. As more information is gained about the high atmospheres of other planets and comparisons are made with our own complex atmosphere, mankind will possess the knowledge to be able to avoid accidentally contaminating the atmospheric envelope of Earth to the point at which some irreversible global imbalance could make this planet uninhabitable. In this context Venus is an important planet for sustained investigation and study because of its many similarities—and many differences—with Earth.

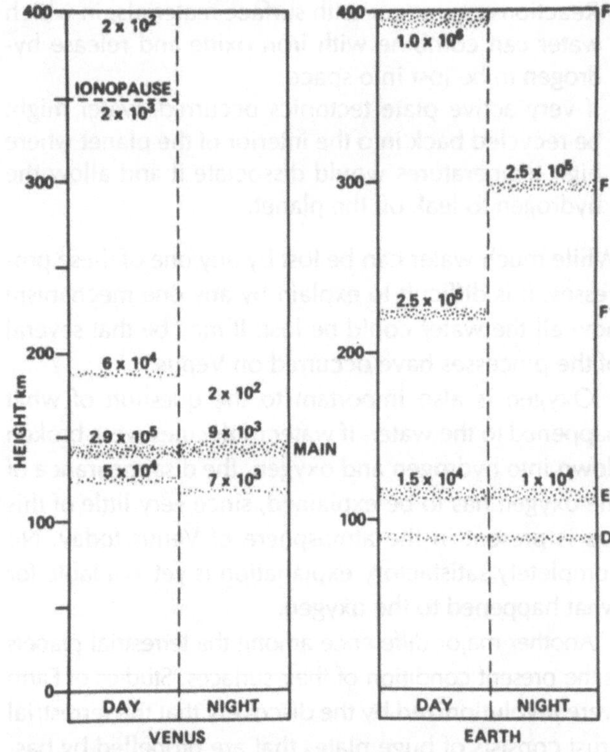

Figure 7.2. The space missions to Venus have allowed comparisons to be made between the ionospheres of Earth and Venus and how they change between day and night.

Scientists had predicted how the upper atmospheres of Mars and Venus might behave. Space probes to these planets showed that these predictions were totally wrong. The result was a rejuvenation of the science of aeronomy and the realization of the importance of mixing between upper and lower atmospheres of a planet. All the planets have extremely complicated atmospheres in which complex photochemical and chemical reactions occur, and interchanges take place between neutral and charged atmospheric gases.

Venus's upper atmosphere, for example, is characterized by lower temperatures at comparative levels than temperatures in Earth's atmosphere, and by unexpectedly large diurnal changes. A mesopause is observed at 117 km (73 mi) where temperature reaches a minimum.

A big question is why the carbon dioxide atmospheres of Mars and Venus should be so stable. It was expected that ultraviolet radiation from the Sun would convert carbon dioxide into carbon monoxide and oxygen. Once formed, carbon monoxide and oxygen should not combine again easily. But there is little carbon monoxide detected on Mars or Venus even though their atmospheres are rich in carbon dioxide. One explanation is that another molecule—incorporating chlorine—on Venus acts as a recombination catalyst.

Most of the water and carbon dioxide that was outgassed from the Earth's interior now resides in the oceans and the carbonate rocks. The nitrogen is still present in the atmosphere. On Mars, the water and much of the carbon dioxide and nitrogen is trapped in cold reservoirs of ice or incorporated into the regolith. Recycling of volatiles through buried sediments may have caused Venus to lose water because of dissociation by heat. Ages ago Mars may have had more carbon dioxide and water vapor in its atmosphere and experienced a strong greenhouse effect that allowed liquid water to flow on its surface. This might be the explanation for the deluge-like floods that seem evidenced by flow features on the present Martian surface. However, unlike the case on Venus, the strong greenhouse was not maintained (because of Mars's greater distance from the Sun) and the planet moved into its present deep-freeze situation. However, because of the ellipticity of the Martian orbit and perturbations to it, there may have been periods of striking climatic changes on Mars analogous to ice ages and warm periods experienced in Earth's past. Layered ice and dust in the water-ice polar caps of Mars lend credence to this hypothesis.

If the luminosity of the Sun was less in the past, Earth could have been protected from a deep freeze by a strong greenhouse effect. Later, by losing its carbon dioxide through biological processes and into carbonate rocks, it was able to avoid a runaway greenhouse as the solar luminosity increased. Venus, by contrast, in the period when solar luminosity was lower, could have had oceans and a milder climate, and could have been sufficiently Earth-like for life to develop and flourish. But as the solar luminosity increased Venus was just that much closer to the Sun for the planet to be pushed into a runaway greenhouse which destroyed its oceans, eradicated its living things, and changed the planet to the present hot, dry global desert.

Most generally accepted models of how the Sun has evolved since its formation require the assumption that the luminosity of the Sun has gradually increased. This leads to some major questions about the evolution of the terrestrial planets. While a lower luminosity in past ages might have allowed Venus to possess an ocean and have a relatively mild climate similar to that of Earth today, the problem remains of explaining how the Earth avoided becoming a deep-freeze planet; it receives only about half the solar radiation received by Venus. The temperature of Earth should have been well below freezing and the whole planet covered with ice shortly after its formation and until about 2.3 billion years ago. However, we know from the records of the rocks that there were oceans on Earth during that time and that life evolved on the planet by 3.5 billion years ago.

The problem with Mars is even more difficult to resolve. Today Mars is a frozen world, yet in times past large quantities of liquid water must have flowed across its surface to sculpt the erosional features seen today. Yet at that time of a lower solar luminosity Mars would be expected to be much colder than it is today.

A suggested way out of this dilemma is to assume that both Earth and Mars had atmospheres of different compositions in the past, atmospheres that were capable of generating a much more efficient greenhouse heating than their atmospheres can accomplish today. Such an early atmosphere might have been one rich in carbon dioxide. Earth, we know, lost considerable amounts of carbon dioxide to rocks and to living things. Mars, too, might be assumed to have lost carbon dioxide when there was plenty of liquid water on its surface.

If Venus had possessed oceans of water early in its history, it might also have had a cool atmosphere. But the question then remains that if Mars and Earth were warmed because of the greenhouse effect of an early carbon-dioxide-rich atmosphere, why did not Venus also have such an atmosphere? If it did, then Venus would be expected to heat up rapidly, despite the low solar luminosity; and as its water entered into its atmosphere, the greenhouse effect would have been strengthened further and quickly the oceans would have been lost to the planet.

If Venus did not have much water early in its history, then the planet would have heated more quickly, and there would not have been time for carbon dioxide to enter into carbonate rock formation. With copious water, as implied by the measurements of the deuterium to hydrogen ratio, carbon dioxide might have been extracted from the planet's atmosphere fast enough to prevent a buildup of a greenhouse effective carbon dioxide atmosphere by outgassing, and Venus might have remained cool for an appreciable time.

Another possibility is that as the solar luminosity increased there might have been a time when Venus's atmosphere contained so much water vapor that much more heat was trapped and the surface temperature was raised high enough to melt surface rocks and deform large ancient impact craters into their present shallow form and to eliminate the smaller impact craters entirely.

As mentioned earlier in this chapter, there are other versions of how the atmospheres of the terrestrial planets might have evolved. According to these interpretations the planets are assumed to have had very large primitive atmospheres somewhat like those of the outer planets and to have lost them by intense solar winds early in the history of the Solar System. In such models, referred to as hybrid models, the present-day atmospheres are the remnants of these large primitive atmospheres that were gathered from the solar nebula at the time of planet formation.

A possible experimental means of resolving some of these questions would be high-resolution inspection of the continents of Venus—particularly Ishtar, which appears to be the oldest highland mass on the planet. If very-high-resolution images can be obtained either from orbit by radar or by cameras from beneath the clouds by means of a winged descent vehicle or a balloon probe, we might be able to find evidence of water flows, as on Mars, or even ancient shore lines which we have not yet identified on Mars.

Scientists now view Venus as a planet that probably started out as a twin of Earth and began to develop in the same way. Life may even have started in ancient oceans. But then came catastrophe. A runaway greenhouse began. Soon the oceans of water had evaporated into the atmosphere and started to leak into space as the incoming solar ultraviolet radiation split the water molecules into their component elements. Venus became the errant twin, throwing much of the incident solar radiation back into space but unable to stave off the inevitable. The promise of life and a habitable planet evaporated like the oceans, and Venus became the hot desert of today.

For the planets the first few hundred million years after their formation set the stage for eons of evolution. By measuring the isotopes of selected elements, planetologists obtain information about the processes during which the material of the planets differentiated. This was the period when heating of the planets allowed light materials to float toward the surface and dense materials to collect toward the center. The terrestrial planets are believed to have metallic cores of somewhat different sizes and chemically distinct crusts of different thicknesses overlying a mantle of iron magnesium silicate.

All the terrestrial planets appear to have been heavily cratered early in their history as a result of a bombardment from space by objects as large as 50 km (30 mi) in diameter (figure 7.3). While evidence of this bombardment has been nearly all removed from the Earth by weathering and plate tectonics, the Apollo expeditions to the Moon and the subsequent analysis of lunar rocks has played an important role in establishing the time of this bombardment. The lunar record places it about 3.9 billion years ago, which was some 700 million years after the formation of the planets. It is a mystery why this intense bombardment took place so long after the planets formed, since it would be expected that the final collection of planetesimals would have taken place much earlier, before the crusts of the planets were established. The intense bombardment took place, however, after the planets had differentiated and had formed crusts.

This differentiation also was responsible for the interior of the planets outgassing volatiles to form the atmospheres of the planets and the hydrosphere of Earth. The bombardment must accordingly have taken place through and modified the primitive atmospheres of the terrestrial planets.

By analogy with the Moon, where the record of evolution seems clearly preserved in the lunar rocks, the

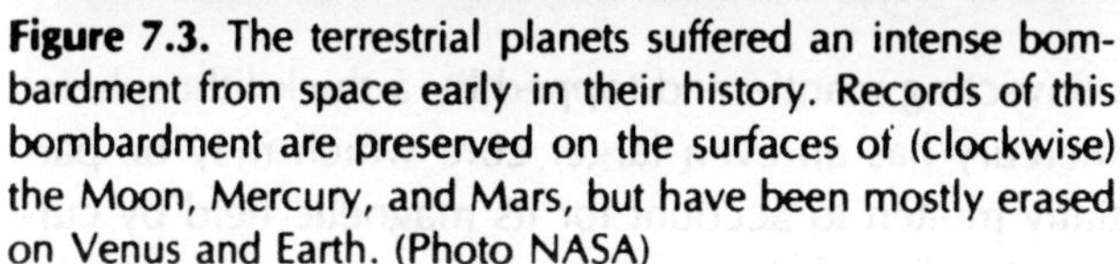

Figure 7.3. The terrestrial planets suffered an intense bombardment from space early in their history. Records of this bombardment are preserved on the surfaces of (clockwise) the Moon, Mercury, and Mars, but have been mostly erased on Venus and Earth. (Photo NASA)

terrestrial planets appear to have heated by the decay of radioactive aluminum 26 so that they differentiated and formed crusts, probably within a few million years of their formation. Later there was a period of igneous activity in which lava flows covered large areas. Then the intense bombardment from space occurred, pockmarking the surfaces with craters of a wide range of sizes. This was followed by subsequent volcanic activity and more lava flows (figure 7.4). Internal heating to the present day within the terrestrial planets probably derives from the radioactive decay of long-lifetime isotopes of uranium, thorium, and potassium. Earth continued with tectonic activity and recycling of crustal material as plates of crust were moved apart by new material flowing from the mantle toward the surface and expanding there from spreading ridges. On Mars we see evidences of all these evolutionary phases, but not of very active tectonics and plate movement because the lithosphere (the rocky material of the planet's outer shell consisting of the crust and the solid outer part of the mantle) became thicker than that of the Earth, which is about 100 km (62 mi) thick. On the Moon and Mercury we see examples of the first episodes, but again there is not much evidence of subsequent plate tectonics, probably because these smaller worlds developed thick lithospheres. Venus is still an enigma mainly because we do not have detailed information about its surface; the Pioneer and Venera radar images do not provide resolution of sufficient detail. Nevertheless, Venus appears to have craters, continents, volcanoes,

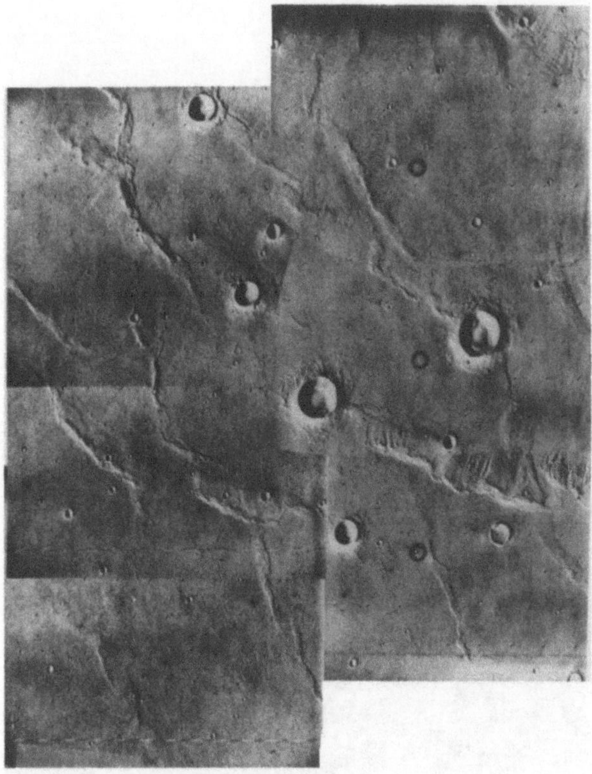

Figure 7.4. Following the bombardment, internal heating of the planets continued and resulted in extensive flows of lava which obliterated many surface features. The flows are clearly seen on (l) the Moon, the (r) Mars. (Photo NASA)

and displays some evidence of folded mountains and lava flows, and canyons which may be rift valleys. The size of Venus approximates that of Earth, and because the temperature gradient in the crust of a planet would be expected to be proportional to the planet's radius, Venus would be expected to have developed a lithosphere somewhat the same as Earth, and plate tectonics would also be expected to develop.

The interiors of the planetary bodies have been investigated directly by seismic studies for the Earth and Moon only, and indirectly by gravitational studies for Mars, Venus, and Mercury. The Earth is known to have four major layers. At the center is a relatively small solid core surrounded by a layer of liquid metallic materials which extends about halfway to the surface. Above this is a mantle of silicates which flow like plastic. This is covered by a surface crust which is of low density and relatively very thin (figure 7.5). The Moon has a very small metallic core and a thick silicate lithosphere. Mars has a core of considerable size surrounded by a slowly convecting mantle and topped by a thick lithosphere. Mercury has an even larger core which may be partially molten to account for its magnetic field by current theories of planetary dynamos.

Venus appears to have a metallic core similar in size to that of the Earth, and the planet probably has an internal structure very much like Earth. Whether the lithosphere of Venus is thick like that of Mars, thin like Earth's, or even thinner than Earth's lithosphere is not presently known. But radar observations from high-resolution orbiting radars or seismic experiments by future Veneras may resolve the question. Also, if there are many large impact craters identified on Venus it will mean that there has not been much tectonic activity on the planet. If craters are rare, as on Earth, then it may be concluded that Venus has been tectonically active like the Earth.

Uplifted plateaus, rift valleys, fault scarps, and ridges seen on the radar images of Venus suggest that there have been tectonic activities on the planet to segment

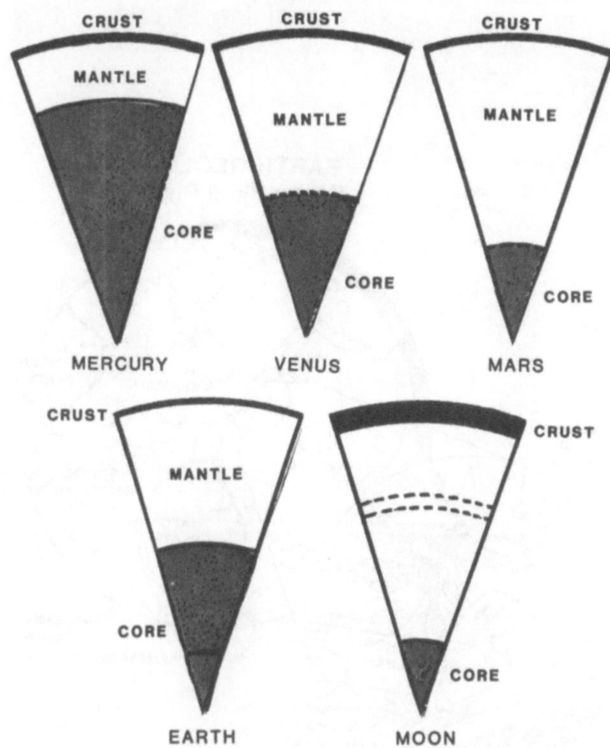

Figure 7.5. All the terrestrial planets have cores, mantles, and crusts which vary in size and may vary in composition. From its size and mass Venus would be expected to have an interior somewhat similar to that of Earth.

the crust. But there is no clear evidence of terrestrial mid-ocean ridges or subduction troughs on the edges of continents—features so characteristic of the global plate tectonics of Earth. Since Venus lacks features like mid-ocean ridges it seems probable that heat is not transferred globally from the interior to the surface by convection as on Earth. The limitation of tectonic uplifts to a few regions suggests instead that there is a vigorous convecting interior from which the heat comes to the surface at several hot spots only (figure 7.6). However, as on Mars, there are volcanic mountains of different sizes and ages, the youngest of which is Beta Regio, associated with a large gravity anomaly. These volcanic areas are probably where the convective hot spots push new crustal material to the surface.

Some Soviet scientists concluded that processes of faulting and subsidence of ancient crust and later flooding by basalts probably occurred relatively recently in geological time. In the highlands province of Beta Regio analysis of gamma spectrometry, as well as mechanical and optical qualities measured by the Veneras, led the Soviets to conclude that the basaltic composition for the loose soil at the Venera 9 site and hard rock at the Venera 10 site showed weathering, with thorium accumulating in the soil. At both sites the impact of the landing capsule raised a dust cloud, as did the landing of Venera 12 in Phoebe Regio, and the Pioneer Venus Day probe in the saddle between Tefnut Mons and Hathor Mons. The concentrations of potassium, thorium, and uranium are quite diverse on Venus compared with the Earth and Moon. In the rolling plains province the surface material is lighter colored and more modified by weathering, as measured by Venera 8. If these rolling plains do represent ancient crust it has a different composition from the crust of the Moon.

Measured wind speeds near the surface of Venus show that fine material generated by chemical weathering can be transported from place to place. The irregular low basins revealed by the radar may thus be partially filled with windblown debris which accumulates and consolidates there like sediments in terrestrial ocean basins.

Another mystery about Venus is the absence of large impact basins or craters. Circular features of diameter greater than 300 km (185 mi) are missing on Venus yet found on other inner planets. It is a possibility that the high surface temperature allows the rocks to deform so that over a long period all the large craters disappear. This would imply that the surface of Venus is old, and that the hot surface conditions have existed for several billion years.

Another consequence of the hot surface is that flows of lava from volcanoes would extend farther than on Earth because the lava would not cool so rapidly. The high atmospheric pressure would also tend to keep volatiles in the flowing lava and make explosive eruptions rare. Eruption centers would thus be expected to produce long single flows building broader and flatter shields than the volcanoes of Earth and Mars.

Perhaps the most interesting difference between Venus and Earth is in the contrast between elevations. Earth

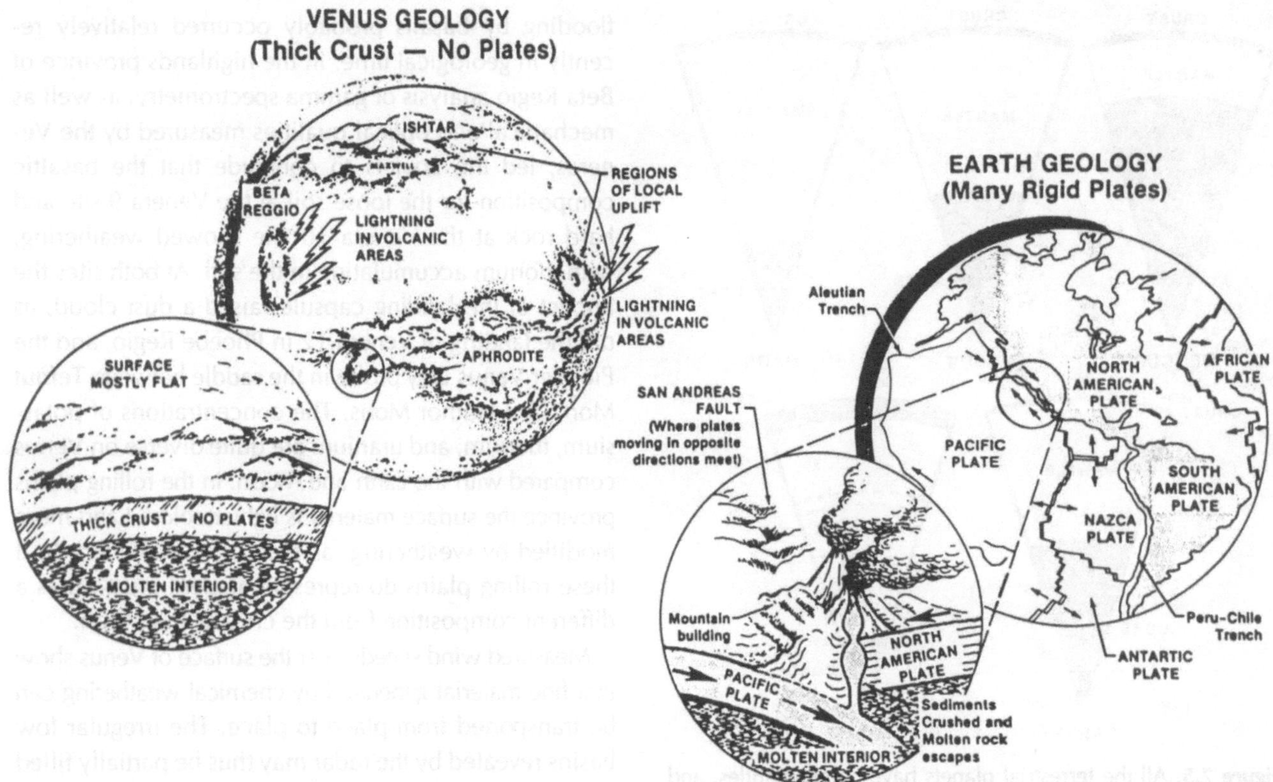

Figure 7.6. From the preliminary radar surveys of Venus it appears that there are major differences between the geology of Earth and Venus. While Earth's internal heat comes to the surface at mid-ocean ridges and spreads plates of crust which subduct back into the mantle, Venus may have a much thicker crust with no moving plates and with the internal heat coming to the surface at specific points in several volcanic areas. (NASA/ARC)

has a bimodal distribution of elevations whereas Venus is unimodal. Earth's topography is sharply divided between the continents and the ocean basins. The topography of Venus is dominated by rolling plains with only a few high and low areas. It is believed that the presence of oceans on Earth makes a large contribution to this bimodal pattern because of the erosion of continents and deposition of the products of erosion around them, coupled with production of low-lying oceanic crust and collection of low-density continental crust into mountainous masses under the influence of plate movement. Material from the high continents of Earth is transported into the deep ocean basins much more slowly than new oceanic crust spreads from the mid-oceanic ridges. This allows the bimodal distribution to continue.

The radar data about the surface of Venus are not consistent with the surface being a deep and porous regolith like that of the Moon, but suggests it is weathered basalt or granite. We are not yet in a position to state whether the surface consists of weathered igneous rocks, as might be implied by the sampling achieved by several of the Venera landers, or of cemented sedimentary rocks which could result from wind transportation of erosional debris. The surface might indeed be a mixture of the two types. Although the Veneras have sampled some rocks at several sites and showed that potassium, uranium, and thorium abundances are more like those in Earth rocks than in Moon rocks, detailed chemical analysis of the surface rocks is still lacking. Such analysis of the rocks and soils of other planets is essential to understanding how their surfaces have been molded to their present forms. Nor do we know how much water is structurally bound within the rocks of

Venus. This is important because the amount of water in the rocky material determines the melting point of rock and its flow characteristics. More water means that rock melts at a lower temperature and flows more easily.

Such water-rich rocks in the lithosphere of Venus would deform easily at the high temperatures of the surface of the planet, and presumably throughout the lithosphere, and high continental masses or mountains could not be supported. If this is so then the high mountainous areas of Venus that we see today must be relatively young formations or must be located over regions where magma continues to rise from the interior to the surface, and Venus is probably still active tectonically.

The rocks of Venus undergo different types of weathering. Chemical weathering would be expected to decompose olivines, pyroxenes, quartz, and feldspars in to magnesite, tremolite, dolomite, and sulfides and sulfates. Mechanical weathering would be expected to disintegrate rocks by spalling and preferential chemical weathering and possibly by wind erosion. Venus is not likely to have the typical mechanical weathering mechanisms of Earth, which rely on water and freezing and biological processes.

Although winds on Venus near the surface do not blow at high velocity they represent the movement of extremely dense air by terrestrial standards, sufficiently dense to move particles up to several millimeter diameter across the surface of Venus. The radar data are, however, inconsistent with Venus being covered by vast areas of windblown debris. However, older windblown sediments in the basins of Venus might have consolidated at the high surface temperatures and not be recognizable as such by radar probing. If so, the low-lying basin floors may not be volcanic lavas like those of Moon, Mars, and Mercury, but layered sediments as mentioned earlier. Conditions in the basins of Venus might even be analogous to the ocean basins of Earth, in which waterborne sediments are replaced by airborne sediments. Such sediments on Venus, as on Earth, would hide from observation the original surface features and their hills and valleys.

The fine detail on the radar maps of Venus is extremely intriguing. Many of the great chasms may be much deeper than shown on the maps because of the relatively low resolution. Also, many peaks may rise much higher than now recorded—again because small very high areas cannot be seen by the radar. The radar images display intriguing wispy streaks, some of which are curvilinear. They are unexplained and may indicate tectonic activity. Some of the bright areas on radar maps may not be slopes, as often interpreted, but fractured zones or areas of sand dunes like those discovered on Mars by the Viking Orbiters. Resolution of the true nature of the surface of Venus really requires optical as well as radar imaging and this will need the development of completely new types of spacecraft able to penetrate below the clouds and return optical images taken during extended flight over the surface of Venus. Aerial mapping of Venus from below the clouds appears essential for a true understanding of the geology of the planet.

Because the atmosphere of Venus has less argon 40 than Earth's atmosphere, it may be that Venus has not been as tectonically active as Earth. Argon 40, formed from the radioactive decay of potassium 40, requires tectonics or volcanics to bring it to the surface, where it can enter the atmosphere. Since the abundance of potassium and the potassium-to-uranium ratio are both Earthlike in the Venus rock samples so far examined, an initial shortage of potassium on Venus cannot be the reason for less argon 40 today. However, there are uncertainties in all the measurements made to date which are great enough to allow for Venus's being as tectonically active as Earth during its evolution. Also, Pioneer Venus detected gravity anomalies which are explainable on the basis of Venus's being tectonically active enough now to support the continental highlands. In addition, lightning detected over what the radar images suggest are young active volcanoes would imply that these volcanoes are active today and that the lightning occurs in the volcanic plumes, a common occurrence with terrestrial volcanoes in eruption.

Another unresolved question about Venus has been mentioned several times already. It centers on the circular features some scientists interpret as impact craters, others as volcanic in origin. Most of these big circular features are located in the flat rolling plains. If they are impact craters then these plains must be very ancient, which would imply that the crust of Venus is ancient. However, if the circular features are volcanic—and their irregularities and shallowness suggests that this is so—then the flat rolling plains crust is very young. Unfortunately we still do not have images of the surface of Venus of great enough resolution to differentiate between impact and volcanic features, and this too may have to wait for optical imaging of the surface.

From its size and the abundance of certain isotopes, Venus would be expected to be transporting heat from its interior to its surface in about the same quantities as Earth. On Earth most of the heat flow from the interior is through the generation of new lithosphere at the spreading ridges. On Venus, by contrast, it is thought that the heat from the interior comes to the surface at specific hot spots associated with upwelling convective flow in the mantle, which leads to continued local volcanism over a region of thin lithosphere. On Venus there are also linear zones with pitted hilly terrain that might represent divergent boundaries with a high heat flow and many active volcanoes. If it were not for the difficulty of subduction processes occurring in the more buoyant lithosphere of Venus, these linear zones might become spreading centers like those we have on Earth.

Even to the casual observer, Mercury, Venus, Earth, and Mars differ markedly from each other, although all are within a small range of sizes compared with other objects in the system (figure 7.7). Important clues as to why they have evolved so differently can be obtained from analyzing the gases of their atmospheres. As discussed earlier, one theory of the formation of the Solar System assumes that when the planets first formed they collected their atmospheres directly from the solar nebula, and these atmospheres probably consisted of gases with a composition resembling that of the Sun and of the giant planets Jupiter and Saturn. The theory then assumes that the primitive atmospheres of the terrestrial planets were lost during the early stages of formation of the Solar System because temperatures were higher then as the Sun began to radiate more strongly.

But the consensus now appears to be that the atmospheres possessed by the terrestrial planets today are more likely composed of gases that were originally pre-

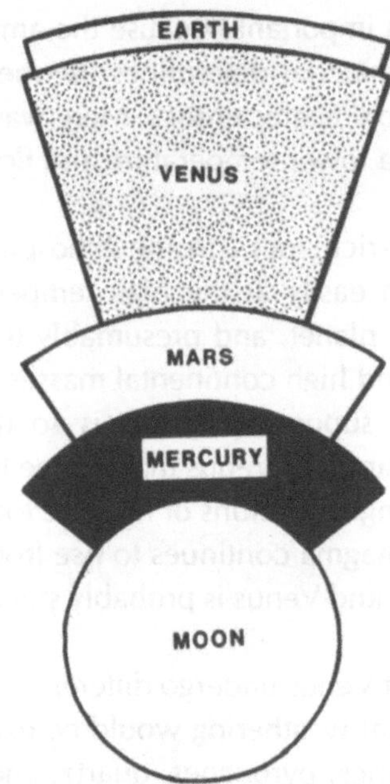

Figure 7.7. The relative sizes of the terrestrial planets are shown in this diagram.

sent as volatile material initially in the solid body of each planet. Within the first few million years after the planets formed, the internal temperature of each planet rose, and associated tectonic activity drove the volatiles from the mantles of the planets; and some of these volatiles remained to form the present atmosphere of each planet. Other volatiles, such as water vapor, condensed or were lost in chemical reactions or into space. On Earth the water formed into oceans. On Mars the water now forms a miles-deep permafrost. On Earth carbon dioxide was converted chemically to carbonate rocks such as limestone and was captured by living things. On Mars and Venus much carbon dioxide remains in the atmosphere to this day.

If this explanation is valid, the amount of each kind of gas now present in the atmosphere of a terrestrial planet should depend mostly on the mass of that planet. The result of missions to Mars showed that this was not so for Mars. Although Mars would be expected to have outgassed smaller quantities of all volatiles than Earth and Venus, it is, compared with Earth, relatively short of carbon, oxygen, nitrogen, and the inert gases neon, krypton, and argon. For some gases the amount is one hundred times less than expected. Following analysis of data from the Viking missions to Mars, scientists suggested that the initial material which Mars obtained from the solar nebula was itself deficient in volatiles compared with Earth. If that were so, smaller quantities of volatiles were available for release from the interior of Mars. Since Earth and Venus are similar in size, mass, and distance from the Sun, scientists also anticipated that, except for water, the amount of volatiles available for release from within these twin planets would be similar, and greater than those on Mars.

Before spacecraft reached Venus, many scientists also agreed that the atmosphere of Venus was 95 to 98 percent carbon dioxide gas, but the rest of the atmosphere was believed to be mainly nitrogen by analogy with Earth. Greatest atmospheric pressure on Earth is about 1 percent of that of Venus, and carbon dioxide forms about 0.03 percent of Earth's atmosphere. The atmosphere of Venus has 300,000 times as much carbon dioxide as does that of the Earth. However, this does not mean that Venus outgassed more carbon dioxide than did Earth. The terrestrial crust contains much carbon dioxide trapped in limestone and some other rocks and in the biosphere and its products. Venus has produced only about twice as much carbon dioxide as the Earth, but the gas has remained in the atmosphere of Venus because there is no abundance of water on Venus to help transform the gas into surface rocks, and most probably there has been no biological use of the carbon dioxide.

A major problem already discussed is how to account for the lack of water on Venus today. Even more important is whether some climatic or man-made change on Earth can increase the amount of carbon dioxide and water in our atmosphere sufficiently for a runaway greenhouse to start. Today we have a much better understanding of the composition and the structure of Venus's atmosphere (figure 7.8) which together with our understanding of Earth's atmosphere may help us guard against man-made irreversible changes to our environment. Carbon dioxide and water trap heat radiation, so that an increase in their concentration would increase the temperature of Earth's atmosphere. In turn this would cause more carbon dioxide and water to

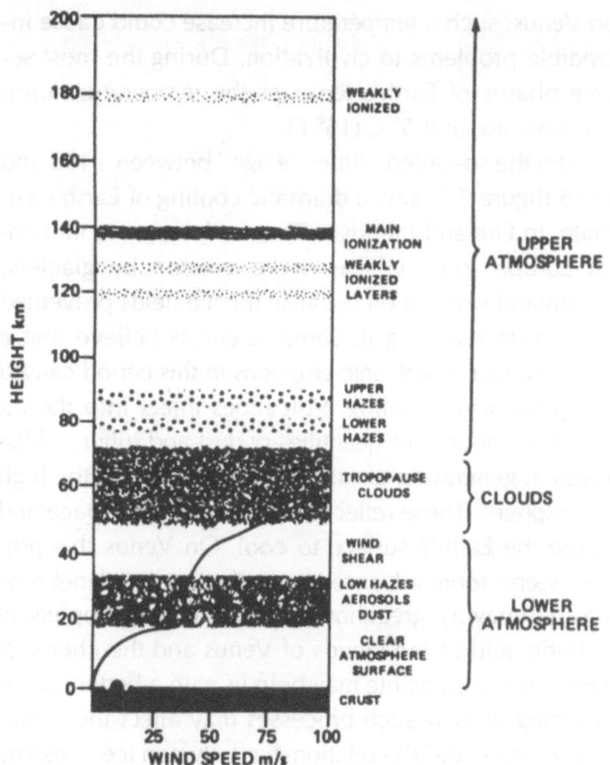

Figure 7.8. The atmosphere of Venus has three distinct regions with quite different characteristics, and these differ markedly from the terrestrial atmosphere and its regions. (NASA/ARC)

enter the atmosphere, with further increases in temperature. Earth's atmosphere might be converted into an atmosphere like that of Venus with all available carbon dioxide in the atmosphere and a surface temperature of about 425° C (797° F).

The main atmospheric ingredient holding in Venus's heat is carbon dioxide. Eighty years of burning fossil fuels on Earth has increased atmospheric carbon dioxide on our planet by 15 percent. Predictions of increased burning of fossil fuels—coal, oil, wood, natural gas—in the future, coupled with opposition to the building and use of nuclear power plants, suggest that carbon dioxide in Earth's atmosphere could be doubled within 50 years. Researchers point out that resulting heat trapped by carbon dioxide could raise the average terrestrial temperature by 1.5°–4° C (3 to 7° F). Although this is minor compared with the searing heat on Venus, such a temperature increase could cause incredible problems to civilization. During the most severe phases of Earth's ice ages the temperature drop was only about 8.5° C (15° F).

Even the so-called "little ice age" between 1450 and 1615 (figure 7.9) saw a dramatic cooling of Earth's climate. In England the River Thames froze; in continental Europe some villages were overrun by glaciers; Greenland was cut off because the ice fields prevented ships from reaching it. Some scientists believe that a large number of volcanic eruptions in this period caused the planetwide cooling. Volcanoes inject into the atmosphere enormous quantities of dust and sulfur, which result in generation of sulfuric acid aerosols in the high atmosphere. These reflect sunlight back into space and cause the Earth's surface to cool. On Venus this process seems more advanced, even though the planet now has a runaway greenhouse. Studying the droplets of sulfuric acid in the clouds of Venus and the chemical processes responsible may help us gain a better understanding of how such processes may affect the terrestrial climate and the relationship between ice ages and runaway greenhouse situations.

Examples of major terrestrial climatic changes resulting from a 1.5° to 4° C change in average temperature might include catastrophic changes in the amount of rainfall in marginal agricultural areas, such as the wheat growing regions of the U.S., Canada, and Soviet Russia. Such temperature changes might also melt a small but important portion of the polar ice caps—enough to raise ocean levels and cause permanent flooding to many population centers along the coastlines of the world.

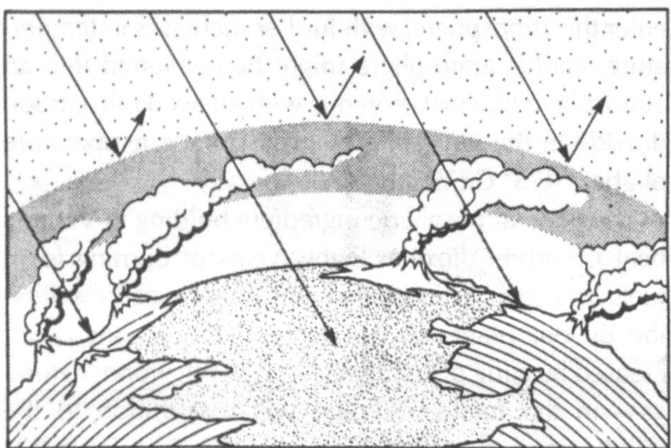

Figure 7.9. There was a "little ice age" on Earth between the years 1450 and 1650, probably caused by series of volcanic eruptions which generated many sulfuric acid aerosols in Earth's atmosphere. These aerosols reflected sunlight back into space and stopped it from heating the Earth's surface. (NASA/ARC)

In 1983 the U.S. Environmental Protection Agency issued a report warning of major changes to Earth's temperature starting before 2000. The trend toward a greenhouse is imminent and inevitable, even though there is still uncertainty about the speed and size of the temperature changes. By 2100 the increase in temperature could be as much as 5° C (9° F), the report said. Such an increase in the average temperature would be accompanied by an increase of three times that amount in the polar regions of Earth. This would cause rapid melting of the polar ice caps, with an increase in mean sea level of as much as 2 m (7 ft). Another report by the National Science Foundation confirmed that a significant terrestrial warming in the next century probably cannot be avoided.

Having calculated atmospheric heating for Venus, an extreme case, scientists expect that calculations of the effects of carbon dioxide on Earth's atmospheric heat budget will be facilitated, and a better understanding may be obtained of whether or not there is a serious threat to Earth.

While the predictions of carbon dioxide heating on Earth might be overstated, it is important to find out as quickly as possible if they are or not. Fortunately, only about half of the carbon dioxide sent into Earth's atmosphere by the industrial revolution and its aftermath has stayed there. Much has been absorbed into the oceans and some taken up by biological systems. Today, Earth's greenhouse effect heats the terrestrial surface about 30° C (55° F), whereas on Venus the effect increases the temperature by 440° C (800° F). On Venus water vapor, sulfur dioxide, the clouds and the high haze assist carbon dioxide to trap solar heat. Amounts of heat taken up by the key absorbers are, carbon dioxide, 55 percent; water vapor, 25 percent; clouds and hazes, 15 percent; and sulfur dioxide, 5 percent. This is based on an atmospheric composition of carbon dioxide, 96 percent; water vapor, 50 parts per million (ppm); sulfur dioxide, 200 ppm; and on cloud thicknesses of 30 km (18.6 mi) for the cloud deck and 12.5 km (7.7 mi) for the haze layer.

Mass spectrometers and gas chromatographs carried into the atmosphere of Venus have identified atmospheric constituents and confirmed that 96 percent of the atmosphere of Venus is carbon dioxide and 4 percent is nitrogen. For the surface pressure of Venus which is 94.5 times that of Earth and for the temperature of 460° C (860° F), these percentages mean that Venus outgassed 1.8 times as much carbon dioxide and 2.3 to 4 times as much nitrogen as Earth, almost as expected.

The instruments were able to detect minor constituents of the atmosphere present to only a few parts per billion (ppb). The results showed that the rest of the volatiles of Venus's atmosphere are quite different from what was expected. There are two types of argon isotopes of significance. Argon 40, the most abundant kind of argon in Earth's atmosphere, is produced by radioactive decay of potassium. Its abundance tells us much about the primitive concentration of potassium and outgassing conditions throughout the history of the planet. Argon 36 and argon 38 isotopes are primordial gases. They provide information about the initial volatile content of planetary interiors and subsequent outgassing. Based on the amounts of carbon and nitrogen in the atmosphere of Venus, about as much argon 36 and 38 was to be expected in the atmosphere of Venus as in the atmosphere of Earth. Instead, concentrations of argon 40 and primordial argon were about equal. About 30 to 70 atoms in every million atmospheric molecules were argon 36. Although the atmosphere of Venus contains about 75 times as many molecules of argon 36 as Earth, the ratio of argon 38 to argon 36 is almost identical to the terrestrial ratio, which means that Venus has many times more of the primordial argon than has Earth.

Another primordial gas, neon, confirms the argon results. The abundance of neon is between about 4 and 13 ppm—compared with 18.2 ppm for Earth's less massive atmosphere. There is about 45 times as much neon on Venus as on Earth. Also, the ratio of neon 22 isotope to neon 20 is 0.07, which is lower than the 0.1 for Earth and close to the ratio for the atmosphere of the Sun.

It appears from these results that Venus, Earth, and Mars could not have consisted initially of materials containing the same proportions of volatiles. Although Venus was provided with about twice as much carbon dioxide and nitrogen as Earth, it received about 50 to 100 times as much neon and primordial argon.

One possible explanation is that the planets were formed from dust grains in the solar nebula which were surrounded by gas at a pressure that diminished rapidly with increasing distance from the center of the nebula. Reactive volatiles such as carbon, nitrogen, and oxygen would be chemically combined within the grains. Rare gases would be adsorbed from the surrounding gas in amounts depending on the pressure in the nebula. In such a scenario the grains forming Mars, Earth, and Venus would have about the same reactive volatile content, but the rare gas concentration would decrease rapidly with increasing distance from the Sun. For this scenario to work, the gas temperature of the nebula would have to be relatively constant, and early outgassing from Mars would need to be some 20 times less efficient than from the other two planets.

There was another surprising result from the mass spectrometers. Although the atmosphere of Venus contains a large excess of neon and primordial argon, the abundance of krypton is only about 3 times larger, and there is less than 30 times more xenon. If the planets accreted from grains it is difficult to explain why one rare gas should enrich the planet more than another. From Mars to Earth the enrichment decreases from a factor of about 220 for neon through 165 for argon, 110 for krypton, to 30 for xenon.

Two important inert gases are produced by radioactive processes. Argon 40 has already been discussed. The other is helium 4, produced when heavy elements such as uranium decay. Argon 40 and argon 36 are about equal in abundance on Venus. On Earth argon 40 is about 400 times as abundant as argon 36. Since there is 75 times as much argon 36 on Venus as on Earth, there is only about one-fourth as much argon 40 on Venus as on Earth. Venus either started with considerably less potassium than Earth or argon is coming from Venus's interior more slowly than it does from Earth's interior. Less tectonic action, a thicker and relatively plastic unfractured lithosphere, and the absence of surface erosion on Venus may be responsible for a slow escape of gases from the interior of the planet.

A measurement of helium in the upper atmosphere of Venus and extrapolation to the lower atmosphere suggests that there are about 12 helium atoms per million molecules. The abundance of helium on Venus is 250 times greater than on Earth. But Venus has not necessarily vented that much more. The present amount of helium on Earth would be produced by radioactivity in the interior in about one million years. Earth's atmosphere is losing helium at a prodigious rate. Probably 5 to 10 times as much helium has been produced and has escaped from the atmosphere of Earth compared with Venus. The present release of gas from the interior of Venus may be slow enough to account for the difference between the radiogenic gas differences on the two planets today if the planets contain equivalent amounts of potassium and uranium.

Among the several theories designed to explain the abundance of argon 36 is the suggestion that Venus had an unusually efficient way of trapping the rare gases during its formation. If so, Venus would be expected to have 75 times as much krypton and xenon as Earth. Instead, Venus has only 3 times as much krypton and xenon. It also has 700 times as much primordial argon 36 as it has krypton. This compares with an argon/krypton ratio for Earth and Mars of only 45 to 1. By comparison the solar atmosphere has 3000 times as much argon as krypton.

A possibility is that Mars formed much earlier than Earth, and Earth much earlier than Venus. Mars may have originated early enough to retain highly radioactive substances such as aluminum 26, left over from a nearby supernova explosion which may have triggered the condensation of the Solar System from the solar nebula. Heat produced by radioactive decay of this aluminum isotope might have quickly driven off many of the volatiles of Mars.

The krypton deficiency mentioned earlier may, however, indicate a far larger contribution by the Sun to Venus's atmosphere than to Earth's during the evolution of the Solar System. The argon/krypton ratio becomes more like the solar atmosphere the closer a planet is to the Sun. This suggests that the material which accreted to form the planets was exposed to a strong flow of gas from the Sun when the planets were forming. Thus, grains and small bodies from which planets were formed would have contained volatiles from the Sun in addition to those from the nebular gas. The material forming Venus may have received more of the solar gases than Earth and Mars. Indeed, Venutian material might have blocked the flow of the solar gas from the material that would form the other two planets.

The material which blew out from the forming Sun during this relatively short period would have struck the mass of material condensing around the orbit of what was to become Venus. In its turn the material would have obstructed the enriched solar wind and prevented it from contributing any significant amounts of rare gases to the materials that would later form Earth and Mars. Hence today we find Earth relatively short of krypton, xenon, and argon.

That Venus's atmosphere has far less krypton than was expected—less than the Sun but more than the Earth—provides more evidence about the formation of the Solar System. It appears that Venus must have received large inputs of various gases during a half-million years or so when the solar wind was much denser than it is today. The theory is strengthened by earlier findings already discussed—for example, the amount of primordial argon 36 on Venus which is 75 times that of Earth. Since Venus has far more of all three of these rare gases—krypton, xenon, and argon 36—than Earth, and has them in proportions more like those of the Sun than of Earth, there is strong likelihood that the gases came from the Sun. This provides some significant evidence about conditions in the early Solar System, and suggests the existence then of a very strong solar wind.

The outcome of the exploration of Venus is that we have found once again that for convenience's sake, human theories often oversimplify things when solving problems of observation. But we have no guarantee that the universe is either simple or constructed in a way to suit the convenience of the human senses and brain patterns. Major mysteries of Venus have been solved, but in their place we now have new and deeper questions which unfortunately we do not seem to have the inclination to assign sufficient human resources to answer. Venus still needs to be explored in greater depth bv orbital high-resolution radar, and by more sophisticated atmosphere probes and surface landers than heretofore. Additionally we need machines that can survey the planet optically on a global scale from beneath the cloud layers. This can be accomplished in part by balloons but preferably by remotely controlled winged vehicles. To probe into the interior of Venus we need a seismic net on the planet's surface the realization of which offers a tremendous technical challenge.

Trying to write the scientific version of how Earth and its neighbor planets formed is philosophically just as important as trying to determine the structure of matter or the state of the universe—probably more so, because it can affect our future. However, during most of the present century, this task has been relatively neglected by scientific investigators. Evolution of the Solar System and of the Earth is the modern version of the story of creation that has always fascinated human civilizations. We have a moral and intellectual duty to write this story in modern terms on the basis of a greater understanding of the physical universe.

Unfortunately this is not regarded as a cost-effective activity, since it brings no immediate dollar return. So it receives one of the lowest of our priorities. Yet it is basic that when searching for causes we also give ourselves the capability of predicting the future and, even more important, perhaps of controlling that future.

In 1976 a report to the NASA Administrator stated that an intensive study of the atmosphere of Venus would begin with the Pioneer Venus mission of 1978. This should be followed, said the report, with an orbital radar mapper mission in the 1980s to image the surface in detail. Surface lander and sample return missions should occur in the 1990s and early 2000s.

So far only the radar mapping mission is receiving attention, and its attempts to gain funding from the federal government have had a somewhat stormy passage. Despite the enormous success of the Pioneer Venus mission, the nation has no firmly defined plans to try to overtake the Soviets in the exploration of Venus. Yet basic science is always the precursor of applied technology. Without basic research the banking accounts of innovation and advanced technology run rapidly into the red, but we should not have to rely upon technology transfer from other nations to bring us back to solvency.

Venus has been unveiled, yet the complexity of the planet has by no means been untangled. The errant twin of Earth is out there, swinging between us and the Sun every 584 days, and challenging us to find out why the "Mistress of the Heavens" is so different from the blue-green planet on which we have evolved.

Professor Lowell made the point about Venus very clearly in his book *The Evolution of Worlds* (1909):

> She stands next us in place, closest in condition and constitution in the primal nebula. Yet in her present state she could hardly be more diverse. This shows us how dangerous it is to dogmatize upon what can or cannot be, and how enlightening beyond expectation often is prolonged and systematic study of the facts.

We desperately need more facts about Venus. Only when we possess accurate information about the interiors, surfaces, and atmospheres of Venus—and all the terrestrial planets for that matter—can a new level of speculation be ended and a new set of questions be based on firm facts. Perhaps we shall never gain all the answers we need to satisfy our insatiable curiosity as we try to find out why we are here and where we are going, and why is Earth, of all the planets of our Solar System, the only one on which life has been encouraged. The answer may have been given by Agnes M. Clerke over 90 years ago in a popular book on astronomy: "Why this has been ordained we are unable to conjecture; we must wait to know."

And in the meantime it would seem that we are morally obligated to use our intellects to attempt to understand our purpose rather than to try to destroy ourselves intentionally or unintentionally. It seems important that we should not forget that beyond Venus there is the challenge of a whole universe waiting to be explored and understood.

APPENDIX
MISSIONS TO VENUS

Venera 1 (USSR): A flyby spacecraft, launched February 12, 1961, Venera 1 passed within 100,000 km (62,000 mi) of Venus in May 1961. Communications with this first interplanetary spacecraft failed en route and no scientific data were returned about interplanetary space or the planet.

Mariner 2 (USA): A flyby spacecraft, launched August 27, 1962, Mariner's closest approach to Venus of 34,833 km (21,645 mi) took place on December 14, 1962. Mariner 2 discovered that Venus is blanketed by cold dense clouds in its upper atmosphere, the surface temperature averages 426° C (798° F) on both day and night hemispheres, and the planet has virtually no magnetic field and no radiation belts.

Venera 2 (USSR): A flyby spacecraft, launched November 12, 1965, Venera 2 passed within 24,000 km (14,900 miles) of Venus on February 27, 1966. Because of a communication failure, an attempt to photograph Venus during flyby was not successful.

Venera 3 (USSR): A hard lander spacecraft, launched November 16, 1965, Venera 3 entered the atmosphere of Venus on March 1, 1966. No scientific data were returned because of a communication failure.

Venera 4 (USSR): A hard lander spacecraft, launched June 12, 1967, Venera 4 entered the atmosphere of Venus on October 18, 1967, and returned data during a 94-minute descent toward the surface. The first spacecraft to enter another planet's atmosphere, its instruments measured atmospheric temperatures from 40° to 280° C (104° to 536° F), and pressures from 0.7 to 20 atm. It determined that the atmosphere is mainly (90%) carbon dioxide. No radiation belts were detected and no significant magnetic field was measured, confirming the Mariner 2 findings. Oxygen and water vapor were both detected in the atmosphere, and a limit was placed on the amount of nitrogen and rare gases that might be present.

Mariner 5 (USA): A flyby spacecraft, launched June 14, 1967, Mariner 5 passed Venus at 3391 km (2105 mi) on October 19, 1967. Radio occultation experiments provided temperature and pressure profiles extrapolated to 527° C (981° F) and 100 atm. at the surface. This spacecraft established the detailed structure of Venus's ionosphere.

Venera 5 (USSR): A hard lander spacecraft, launched January 5, 1969, Venera 5's bus entered the atmosphere and was destroyed. A descent capsule entered the nightside atmosphere on May 16, 1969. It transmitted data during 53 minutes of descent, including measurements of temperature, pressure, and atmospheric composition.

Venera 6 (USSR): A hard lander spacecraft, Venera 6 was launched January 10, 1969. The bus entered the atmosphere and was destroyed. The landing capsule entered the nightside atmosphere on May 17, 1969, and transmitted data during 51 minutes of descent. Instruments carried by the lander determined that water vapor content of the atmosphere is low. Some nitrogen was detected, but nitrogen and measured carbon dioxide constituted 93 to 97 percent of the atmosphere; and the oxygen content is less than 0.4 percent. Extrapolation of data to the surface indicated a surface temperature of 276° C (528° F) and a pressure of 90 atm.

Venera 7 (USSR): A soft lander spacecraft, Venera 7 was launched August 17, 1970. The bus entered the atmosphere and was destroyed. The capsule penetrated the nightside atmosphere on December 15, 1970, and data were transmitted during the descent and for 23 minutes from the surface. Measurements indicated a surface temperature of 543° C (1009° F) and a pressure of 90 atm. at 5° S latitude and 351° longitude.

Venera 8 (USSR): A soft lander spacecraft, Venera 8 was launched March 27, 1972. The bus was destroyed on entering the atmosphere. The lander reached the dayside surface of Venus on July 22, 1972 at 10° S latitude and 335° longitude and transmitted data from there for 50 minutes. This mission determined the amount of uranium, thorium, and potassium in the surface materials sampled at the landing site, and showed that these elements were present in proportions greater than in terrestrial rocks. The lander's instruments measured a surface temperature of 530° C (986° F).

Mariner 10 (USA): A spacecraft bound for Mercury which passed Venus en route. Launched November 3, 1973, the spacecraft flew past Venus at 5793 km (3600 mi) on February 5, 1974, and obtained first pictures of Venus from a spacecraft. They revealed the structural details of the clouds in ultraviolet light. Confirmed the reality of the C-, Y-, and psi-shaped markings, and confirmed the four-day rotation period of the ultraviolet markings. It found significant amounts of helium and hydrogen in the upper atmosphere, and photographed high-altitude haze layers in the upper atmosphere above the cloud tops. It confirmed Venus's lack of a magnetic field of any consequence, determined the detailed structure of the ionosphere, and established temperature and pressure profiles into the high atmosphere.

Venera 9 (USSR): An orbital and soft lander spacecraft, Venera 9 was launched June 8, 1975. The orbiter entered orbit around Venus from which it surveyed the planet. Orbital elements were: period, 48 hours 18 minutes; inclination, 34° 10′; periapsis, 1620 km (1007 mi); and apoapsis, 66,614 km (41,392 mi). The lander capsule reached the dayside surface on October 22, 1975 at 33° N latitude, 293° longitude. It returned first picture (black and white) from the surface of Venus, and transmitted science data for 53 minutes from the surface, and measured wind speeds, pressure, temperature, and solar radiation flux throughout the atmosphere to the surface.

Venera 10 (USSR): An orbiter and a lander spacecraft, Venera 10 was launched June 14, 1975. The lander capsule reached the surface on October 25, 1975 at 15° N latitude, 285° longitude. It returned data from the surface for 65 minutes, including a picture of the landing site, and repeated experiments of Venera 9 throughout the atmosphere. Orbiter surveyed the planet and looked at the surface with bistatic radar. It determined that surface elevation differed by only a few kilometers along the track of the orbiter.

Pioneer Venus Orbiter (USA): Launched on May 20, 1978, Pioneer Venus Orbiter arrived at Venus and was inserted into orbit on December 4, 1978. From there it surveyed the planet with a battery of science instruments for many years. It discovered large continental areas on Venus and radar mapped a large part of the planet's surface. It gathered much data about the planet's ionosphere and interaction with the solar wind. The nominal mission was completed on August 4, 1979, but the extended mission continued into the mid-1980s with decay of the orbit and entry into the Venus atmosphere expected in early 1992.

Pioneer Venus Multiprobe (USA): Launched August 8, 1978, the Multiprobe bus carried four probe spacecraft—one large and three small probes—to penetrate Venus's atmosphere and gather data at widely separated global locations. All probes were successful in entering the atmosphere on December 9, 1978 and one transmitted data for a short while from the surface. Much new information was obtained about the atmosphere from high altitudes down to the surface. The bus also gathered data in the upper atmosphere before burning up.

Venera 11 (USSR): A Soviet lander spacecraft, Venera 11 was launched September 9, 1978. The capsule reached the surface of Venus on December 25 at 14° S latitude, and 299° longitude. It measured temperature, pressure, wind velocity, and composition. Also, it investigated the scattered radiation from the Sun and monitored thunderstorm activity. Attempts to transmit images of the surface were unsuccessful. The spacecraft which transported the lander to Venus flew by the planet at 25,000 km (15,535 mi).

Venera 12 (USSR): A similar vehicle combination to Venera 11 was launched 12, 1978. The Venera 12 capsule reached the surface on December 21 and made the same types of measurements as Venera 11 but at a location of 7° S latitude and 294° longitude. Scientific data, but no images, were obtained. The spacecraft which carried the lander to Venus continued on to fly past the planet at 25,000 km (15,535 mi) and gather science data as it did so.

Venera 13 (USSR): An orbiter and lander combination spacecraft, Venera 13 was launched October 30, 1981 and arrived at Venus on March 1, 1982. The mother spacecraft continued in a flyby trajectory but acted as a communications relay for the lander. The lander capsule reached the surface, from where it sent data for about two hours, recorded a temperature of 457° C (855° F) at the landing site, transmitted the first color pictures of that surface (eight pictures in total), and scooped up a soil sample. Analysis of the rocks showed a composition similar to terrestrial volcanic rocks.

Venera 14 (USSR): An orbiter and lander combination spacecraft launched November 1, 1981, Venera 14 arrived at Venus on March 5, 1982. The mother spacecraft followed a flyby trajectory as with Venera 13. The lander of this mission also successfully reached the surface and transmitted high-resolution pictures from a site about 960 km (600 mi) east of Venera 13. Transmissions continued for about one hour. The pictures showed many flat rocks, areas of soil, and areas swept clean of soil, probably by winds.

Venera 15 (USSR): Launched on June 2, 1983, Venera 15 entered polar orbit of Venus on October 10, 1983. Six days later the spacecraft began to map the surface with high-resolution radar, revealing impact craters, mountains, and other topographic details to a resolution approaching two kilometers—ten times better than the Pioneer Venus images.

Venera 16 (USSR): On June 7, 1983, the Venera 16 polar orbiter was launched successfully and arrived at Venus in October 1983. The spacecraft began operations in orbit on October 20, and its radar mapping of Venus confirmed the presence of features resulting from major tectonic processes. A continent about twice the size of Australia was reported in the north polar region.

Future missions to Venus planned through the 1980s include an American Venus Orbiter Radar Mapper and several Soviet spacecraft, including VEGA (launched in December 1984), which will place balloon probes in Venus's atmosphere for long-term observations within that atmosphere.

Index

Bei Fragen zur Produktsicherheit wenden Sie sich bitte an:
If you have any questions regarding product safety,
please contact:

Walter de Gruyter GmbH
Genthiner Straße 13
10785 Berlin
productsafety@degruyterbrill.com